Emotional Reasoning
Insight into the Conscious Experience

Eva Déli

Nyiregyhaza, Benczur ter, Hungary

CRC Press
Taylor & Francis Group
Boca Raton   London   New York

CRC Press is an imprint of the
Taylor & Francis Group, an **informa** business

A SCIENCE PUBLISHERS BOOK

First edition published 2025
by CRC Press
2385 NW Executive Center Drive, Suite 320, Boca Raton FL 33431

and by CRC Press
4 Park Square, Milton Park, Abingdon, Oxon, OX14 4RN

CRC Press is an imprint of Taylor & Francis Group, LLC

Library of Congress Cataloging-in-Publication Data (applied for)

ISBN: 978-1-032-54444-1 (hbk)
ISBN: 978-1-032-54447-2 (pbk)
ISBN: 978-1-003-42491-8 (ebk)

DOI: 10.1201/9781003424918

Typeset in Palatino Linotype
by Prime Publishing Services

To Sue: "For I was hungry, and you gave me something to eat, I was thirsty, and you gave me something to drink, I was a stranger, and you invited me." Matthew 25–35

Preface

"It is characteristic of mankind to make as little adjustment as possible in customary ways in the face of new conditions."

— Robert and Helen Lynd

The United States Declaration of Independence was a revolutionary document to assert people's rights to pursue happiness. The world had to wait over two centuries for the scientific study of joy and fulfillment. Emotional Intelligence has uncovered a profound link between happiness, health, and well-being. This fusion of emotion and intellect signifies a significant shift in our understanding of the human psyche, as positive psychology highlights the importance of emotions in the family, workplace, and even law enforcement.

In an age where technological revolution promises astonishing transformations in work and social life, it becomes imperative to understand the nuances of human cognition and emotion. Artificial Intelligence transforms the workplace, liberating us from many tasks with the potential to make work exciting, challenging, and humane. We are at a crucial moment when focusing on intellect requires capitalizing on the diverse cultural, racial, and sexual human potential.

This book, *Emotional Reasoning: Insight into the Conscious Experience*, ventures into the intricate realms of emotions and consciousness, juxtaposing scientific explorations with philosophical discourses.

The human mind is a complex and mysterious entity shaped by millions of years of evolution. While we celebrate its logic, creativity, and emotional capabilities, it also harbors quirky, baffling, and weird qualities. In a radical departure from traditional neuroscience, this book builds on this strange duality to search for answers for emotions and consciousness in the realms of physics.

This journey creates a new foundation for human motivation, which might serve as the foundation of future society. The underlying thesis is that emotionally secure and content individuals are intrinsically motivated to work toward the greater good. Productive citizens will actively chart the direction of the increasingly faster-paced technological progress.

Environmental factors, such as residence, parents' occupation, and income, influence a child's IQ.[1] The essential role of the home in personal development leads to intergenerational stability of beliefs. According to some Biblical interpretations, the Israelites could not enter the Promised Land with a slave's mind. The psychological renewal took forty years of desert wandering because the older generation had to die off. Can social transformation happen at a faster pace? This book offers a fast track for societal evolution.

Personal skills (so-called "soft skills") can dramatically improve the social environments. Although changing the mind and mastering emotions is possible at any age, teaching social skills can start in Kindergarten. Although feeling permanent, our mindset changes constantly as we try to keep up with our environments. Career changes, adult education, and social changes continuously update our understanding. These skills can transform societal interactions by fostering fairness, honesty, and collaboration. The book underscores the idea that physical exertion can increase physical fitness, and overcoming mental challenges can pave the way for wisdom and contentment.

For example, at the turn of the twentieth century, cars began to clog the streets and cities. Cities responded by setting up roles to regulate traffic. Yet, initially, drivers resisted the need to obey these laws. Nevertheless, we cannot imagine our congested roadways and fast-paced traffic without following traffic rules. This book introduces an operational understanding of emotions. It supplies internal motivation for change, with the potential to fundamentally transform our social fabric.

In the intricate tapestry of human experience, emotions and consciousness weave the threads that define our existence. From the gentle caress of joy to the depths of sorrow's abyss, from the spark of curiosity to the fire of passion, our emotions color our lives, shaping our perceptions and driving our actions. At the heart of it all lies consciousness, that enigmatic beacon of awareness that illuminates the landscape of our minds.

The tome stands as a testament to the prescience of the Declaration of Independence. It emphasizes the democratization of well-being as a bulwark against societal discord. "Emotional Reasoning" is not just a book; it's a journey. A journey that acquaints readers with the intricate workings of their minds and provides practical tools for self-betterment.

---

[1] Makharia et al., 2016.

As you progress chapter by chapter, you'll be introduced to foundational concepts, delve into the intricacies of emotions, explore their role in performance, and culminate in practical strategies to enhance well-being. Each chapter ties back to the central theme–pursuing a meaningful life, colored by emotions and illuminated by consciousness.

In this odyssey of self-discovery and growth, prepare to challenge long-held beliefs, embrace novel perspectives, and embark on a transformative journey to understand the very essence of human existence. The end goal? A life of purpose, contentment, and profound happiness.

Acknowledgments

To Sue for taking me in and supporting me during the most challenging period of my life.

To Ellie Amundson for her suggestions on self-editing my book.

To Larry Arnst for many valuable suggestions and for believing in my work.

To Dr. Mark O'Malley for the painful editing of the manuscript.

To Chat GPT for helping me with a search for information.

Contents

Introduction

The very essence of science lies in expanding and challenging the established norms and boundaries of our understanding. Still, proposing a novel hypothesis is a bold undertaking, fraught with uncertainty and risks. Questioning long-held beliefs often triggers resentment and skepticism, if not outright mockery. However, new ideas should be evaluated on their ability to solve pressing conflicts and simplify fundamental problems.

A retrospective journey through the annals of scientific history reveals a path dotted with uncertainties and groundbreaking revelations. Though seminal in their contributions, figures like Newton weren't exempt from misadventures in their quest for knowledge. Though postulated by respected physicists, current theories, such as the multiverse concept, sometimes challenge foundational principles of physics. Like Newton's occult explorations, the multiverse concept disregards fundamental physical principles, such as energy conservation.

Hence, scientific brilliance does not render someone immune to error. Science's annals are filled with luminaries like Wegener, Mendel, and Darwin, whose pioneering theories, though initially met with skepticism or ridicule, shaped the foundations of their respective fields. Such tales are a reminder of the dangers of rigid thinking and the importance of keeping an open mind in science.

There is an automatic jerk reaction of hostility or outright derision to ideas outside the established norms. Darwin, for example, harbored such apprehension of ridicule that he delayed publishing his findings for decades. Is it possible to do better? I believe so. The merit of a scientific theory should be based on its ability to yield fresh perspectives (Doerig et al., 2022).

This book offers a radical hypothesis to uncover the nature of consciousness that bridges the worlds of neuroscience, psychology, and physics. It suggests that interacting with the environment through the sensory system makes our brains part of its energy-information cycle. This continuous exchange, subject to the laws of physics, might hold the key to unraveling the nature of consciousness. The brain's ability to consume information and build intellect inspires a spiritual wonder and awe at the coherence of reality and our deep connection to the physical universe (Preston et al., 2023).

While animal life-cycles are bound to the changing seasons, yet remain unchanged over millions of years. Most of human existence followed a similar patern, with most people, including virtually all peasants were illiterate, plagued by disease, frequent bouts of starvation, and worries of survival.

Recent technological advancements and societal shifts have drastically influenced our life trajectories. Today, the concept of self-realization and finding purpose is no longer confined to a select few philosophers, scientists, or spiritual leaders. The democratization of knowledge and the onset of the technological era have made it possible for every individual to seek a life of meaning and purpose. The unending flow of information about technical and social changes dictates a fast pace of change in our minds. The inability to keep pace with this rapid change can trigger mental conflict, anxiety, and even depression.

This very idea of purpose and meaning lies at the core of this volume. It posits that updating understanding and purpose inherently goes hand-in-hand with morality and ethics. We are at a threshold of a paradigm change, where we have the tools needed to discover our purpose and role in the world. In the past, only philosophers, religious leaders, and artists had the luxury to reach their full potential. With the spread of automatization and artificial intelligence (AI), living with meaning is becoming possible for everyone.

The availability of tools to achieve personal meaning and satisfaction inspires trust in the system, provides optimism for the future, and dictates a strong sense of community, making actions that harm society or disrupt the social fabric inherently repulsive.

The natural laws governing our interaction with the environment and the role of emotions in our relationships and lives are the deeply seated qualities of our personal universe. The book you keep in your hands offers a guide to this paradigm. When you finish this book, you will understand how the mind operates and have hands-on tools to initiate positive life changes for you and your loved ones. Understanding these vital relationships and subject matter will put you at the forefront of a rising social tide.

This practical guidebook will expand your mind. You may have to reconsider some preconceptions about emotions and behavior. Your action plan (even minor modifications in thinking and behavior) will enhance your contentment and better your life.

I hope this tome serves as a catalyst, sparking curiosity and fostering a sense of emotional equilibrium and purpose. With newfound knowledge and tools, readers will be well-equipped to navigate life's challenges, making meaningful and positive contributions to society. As you delve deeper into the intricacies of the human mind and its interplay with the environment, you'll realize that the pursuit of truth and understanding is not just a scientific endeavor but a deeply personal and transformative journey.

Chapter 1

The Brain

"*The human brain has 100 billion neurons, each neuron connected to 10 thousand other neurons. Sitting on your shoulders is the most complicated object in the known universe.*"

— Michio Kaku

The subject of consciousness has intrigued humanity throughout the ages. The fast progress in neuroscience recently gave rise to the optimistic expectation that the understanding of consciousness was just around the corner. However, the 25 years old bet between a cocky neuroscientist (Christof Koch) and philosopher (David Chalmers) was decided in the philosopher's favor. Neuroscience still has no idea how the brain achieves consciousness, forms emotions, or the exact definition of these concepts.

It is a fascinating enigma of how a mere lump of brain matter can harbor, transmit, and receive vast amounts of information, conjure subjective conscious experiences, and even introspect upon itself. The mind-body problem and the personal nature of experiences present persistent challenges. In the pursuit of identifying the neural correlates of consciousness (NCC), techniques such as functional magnetic resonance imaging (fMRI) and electroencephalography (EEG) serve as windows offering glimpses into the dynamic brain activity during conscious states. The "global workspace" concept proposes that consciousness emerges when information integrates across various brain regions, enabling cohesive awareness.

Altered states of consciousness caused by anesthesia, meditation, or psychedelics can uncover the mechanisms underlying shifts between conscious and unconscious states. The interdisciplinary synergy of advanced brain imaging techniques and AI has become indispensable in grappling with these complex questions. However, our quest to understand consciousness must start with a comprehensive overview of the brain's architecture.

How did the brain, the body's most complicated part, emerge? In the tapestry of evolution, the Planaria might be the first brainy animal and the possible ancestor of the vertebrate brain. This remarkable organism boasts unparalleled regenerative abilities, capable of regrowing an entire body from a mere 1/100th of its original form. Its capacity to regenerate a new head or tail renders it a favored subject in organ and limb regeneration research studies.

The sensory neuron serves as the sentinel, detecting specific stimuli like touch or chemical cues and relaying the information to the interneuron. The interneuron, in turn, processes the sensory input and generates a response signal, which is transmitted to the motor neuron. Finally, the motor neuron sends commands to the muscles for coordinated movement or behavior.

In the narrative of evolution, the subsequent pivotal phase involves the development of the limbic brain, a conglomerate of structures located above the brainstem. In creatures ranging from fishes to reptiles, the superior sensory processing capacity of the limbic brain facilitates the development of finely coordinated behaviors. For instance, agile sharks' ability to sense their prey's minuscule electromagnetic fields transforms them into adept predators.

The culmination of the brain's evolutionary journey resides in the cortex, the epicenter of consciousness and emotions. Complex emotions, the bedrock of intelligent social conduct, memories, learning, and strategic planning, are rooted in this complex web of neural connections. It is to the cortical brain that our focus now turns as we strive to unveil the intricacies and dimensions of consciousness.

The Anatomy of the Brain

"The brain is just the weight of God." — Emily Dickinson

In ancient thought, the philosopher Aristotle postulated that the seat of perception and the soul's self-nourishment must reside within the heart. To him, the brain merely served as a radiator, a mechanism to prevent the heart from overheating. However, the strides of neuroscience have since relegated the heart to the periphery, elevating the brain as the pivotal hub for perception and the orchestration of the soul.

Biological functions such as heart rate and aerobic capacity boast distinct and defined energy requisites. Yet, these demands pale compared to the brain's insatiable appetite, consuming a staggering 20% of the body's oxygen and 25% of its glucose, underscoring the indisputable primacy of cognition. Orchestrated by the intricate interplay of countless neurons, our actions arise from the simple act of walking to the complex

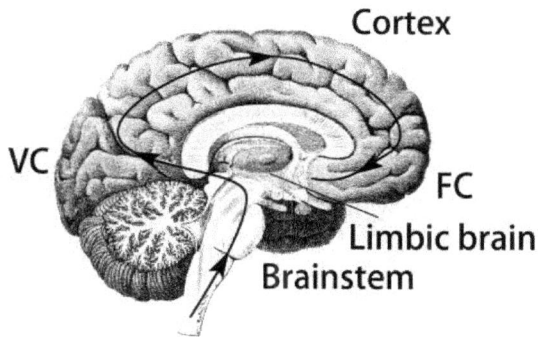

Figure 1. Information flow in the human brain during a stimulus. FC (Frontal cortex), VC (visual cortex). The stimulus arrives in the brainstem and moves to the limbic brain. The cortical processing begins in the sensory cortex and continues in the frontal, associative areas. Response to stimulus coincides with the reversal of the information flow.

realms of artistic creation or primal instincts like 'fight' or 'flight'. Indeed, the brain stands as the supreme command center (as depicted in Figure 1).

Even amidst the cacophony of sensory inputs, the brain makes sense of the outside world with remarkable fluency. This feat is achieved through three intertwined, progressively complex modules operating in a meticulous hierarchy. Each stage of sensory processing builds upon the last, fostering escalating intricacy and nuance.

The initial evaluation of sensory inputs within the brainstem serves as a swift precursor to the subsequent assessments carried out within the limbic system, culminating in cortical perception. The transient activations of the prefrontal cortex act as the alchemical catalysts, conjuring forth the intimate realm of private consciousness.

The Brainstem

Nestled at the brain's foundational core, the brainstem serves as a nexus between the cerebral expanse and the spinal cord (depicted in Figure 1). Analogous to an old switchboard, this ancient region presides over vital functions—posture, respiration, and slumber—comprising the inaugural step in sensory processing. Its extensive interconnections with the cortex, primarily the frontal domain, bestow the ability to swiftly craft overarching sensory impressions and responses inherently influenced by one's mood.

Moreover, the brainstem assumes the role of a pivotal crossroads, where incoming sensory signals traverse from one side of the body to the opposite hemisphere of the brain, only to return as outgoing motor innervations bound for the muscles. These incoming environmental signals undergo a transformative translation, manifesting as synchronized rhythmic patterns known as the brain's oscillations or brain waves, effectively encoding spatial relationships into neural symphonies.

The Cerebellum

Nestled beneath the cerebrum, the cerebellum may constitute merely 10% of the brain's overall weight. However, it houses a staggering 80% of the brain's neurons, dedicated to orchestrating the symphony of muscle movements and maintaining posture. Its specialized sensors are attuned to discern even the subtlest shifts in balance, promptly dispatching signals to establish harmonious connections. Beyond this role as the guardian of physical equilibrium, the cerebellum assumes a more intricate mantle, deftly refining movements through repeated practice. Moreover, it takes on the crucial responsibilities of governing eye movements, facilitating language processing, and weaving the intricate fabric of memory.

The Limbic Brain

Positioned atop the brainstem, the limbic brain forms a cohesive assembly of modular structures (depicted in Figure 1). It serves as a dynamic interface bridging the cerebral cortex's intellectual capacities with the subconscious, autonomic functions anchored within the brainstem. Intriguingly, these limbic modules seamlessly integrate into the cortical regulatory circuits, assuming a pivotal role in orchestrating cognition, emotions, and memory.

Within this intricate network of modules, the amygdala emerges as a critical contributor to the genesis of emotions, particularly evident in processes like fear conditioning. Meanwhile, at the heart of this arrangement lies the hippocampus, acting as the epicenter for the repository of memories and facilitating spatial orientation. Given cognition's profound reliance on an acute environmental awareness, any disruption leading to spatial confusion directly imperils the fabric of one's sense of self. This phenomenon frequently stands as the harbinger, the initial symptom heralding the onset of various mental disorders.

The thalamus is a central information relay and processing station before the cerebral cortex. Input from all sensory systems except the olfactory system is processed and filtered in its two large lobes. Likewise, signals from the basal ganglia and cerebellum are relayed to the motor cortex. The thalamus also generates sleep spindles,[2] contributing to emotion, pain, and memory formation. In addition, it has an essential role in regulating sleep and wakefulness.

---

[2] Sleep spindles are bursts of brain activity during sleep.

The Cortex

The brain's outermost layer, the cortex, is a convoluted gray neural tissue enveloping the limbic structure. Yet, despite its prominence, the unique arrangement of folds, characterized by gyri and sulci, has remained an enigma in the realm of neuroscience. The question persists: What underlying rationale might account for the cerebral wrinkling that defines the cortex?

Central to the cortex's role is its function as an information processing hub, diligently sifting through inputs from the sensory system. This role requires proportionality between cortical volume and the organism's size, leading to a curious relationship. As the brain grows in tandem with an animal's dimensions, the cortical volume, stretched into a thin veneer, undergoes an exponential expansion that far outpaces the brain's increase in volume. In response, the burgeoning cortical tissue folds and crinkles, a strategy to ensure it remains intricately connected to the underlying neural substrates, as highlighted by research led by Im et al. (2018).

In larger animals, the remarkably intricate accordion-like folds within the cortex facilitate the emergence of traveling cortical waves (Pang et al., 2023). The inherent robustness of the cortical architecture molds these waves, invoked by tasks, into expansive brain-wide patterns akin to the undulating crests of ocean waves. These globally synchronized waves ripple through both cortical and subcortical structures in harmony with states of arousal. This orchestration underpins the brain's extraordinary intellectual prowess.

Muscle Twitches: During REM sleep, our bodies are partly paralyzed, but muscle twitches activate the sensorimotor cortex, which is responsible for movement coordination. These meticulous and precise twitches aid in developing a sense of self-awareness and body comprehension. During the gestational period, spontaneous muscle activations set in motion a cascade that unfurls the fetal brain's long-range neuronal connections. Concurrently, the intricate process of selective pruning clears away superfluous neuronal links. Adolescence marks a pivotal phase, refining the synaptic map to mirror individual experiences and the evolving contours of the body.

Unveiling an intricate dance of interactions, the brain's connection map becomes as idiosyncratic as fingerprints, a testament to the relentless optimization of neural circuits through regular engagements. This intricate network orchestrates a seamless translation of neural innervation into impeccably precise muscular movements effortlessly summoned at a moment's notice. However, as the stage of life advances into the realm of aging, a secondary phase of irreversible decline befalls the once-stabilized adult connection map, embodying the transient nature of human existence.

The Brain as a Pinhole Camera (Camera Obscura)

"Who would believe that so small a space could contain the images of the whole universe?" — Leonardo Da Vinci

How does the brain reconstruct the symphony of sounds, the tapestry of scents, the mosaic of images, the choreography of movement, the textures of touch, and even the echoes of pain? The elucidation of these mysteries might draw inspiration from the annals of physics. However, exploring this insight must start in an artist's studio.

Venturing back to the nineteenth century, we encounter the practices of European artists who harnessed the pinhole camera to replicate visual imagery (as depicted in Figure 2). This ingenious contraption, a light-proof box punctuated by a minuscule aperture, could project a diminutive yet inverted depiction of a lit scene onto the box's inner rear surface. Strikingly, this process bears a remarkable semblance to the sensory cortex's inverted portrayal of the surrounding world.

A remarkable dance of sensory data unfolds as sensory data from the left of the body is processed in the right side of the cortex and vice versa. This arrangement extends even to vision, where signals from the right eye journey to the left side of the brain. A symphony of reciprocity unfurls as motor innervation traverses sides en route to the muscles. The complexities do not halt at mere inversion; they extend to processing inputs upside-down. Thus, the toes communicate sensations to the cerebral hemisphere's upper reaches.

At the same time, the head and hands find representation along the cortex's flanks, culminating in the face being relegated to the furthest reaches. This intricately orchestrated symphony persists in outgoing impulses to the motor cortex, adhering to the identical organizational blueprint. In the above organization, the strongest cortical field at the center ensures the cohesion of extremities. Inversely, the innervation of the hands and face at an extreme distance from the center allows fine manipulation.

Within the brainstem, the juncture where neural pathways traverse between the cortex and the body resembles the mechanism of a pinhole camera. As the infant's brain takes tentative steps in deciphering this inverted sensory map, an uncanny mastery over this perceptual inversion is acquired. Remarkably, sensory perception is an involuntary pursuit entirely distinct from conscious intent. We find ourselves powerless in the face of stimuli, unable to halt their reception or our comprehension. It is a journey where evolutionary and intellectual advancement unfolds irrevocably, forever marching forward. Once we unlock the reading skill, decoding the significance of a STOP sign becomes an inexorable consequence.

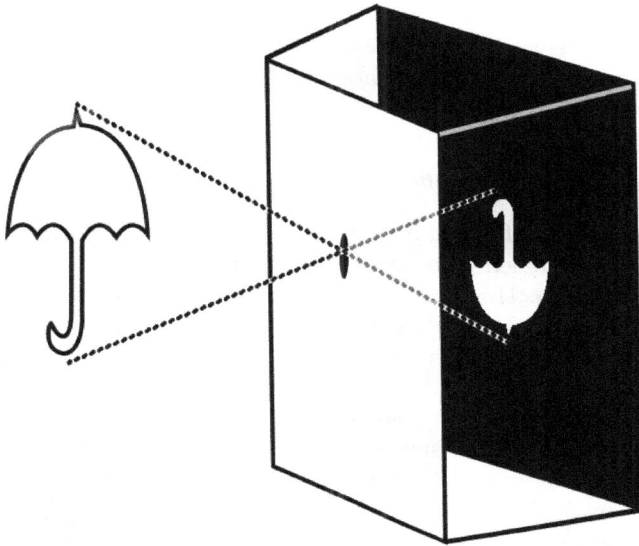

Figure 2. The Pinhole camera. Light only enters the box through a small hole. The umbrella's small projection on the back wall appears upside down with the sides switched.

Intriguingly, it's vital to acknowledge that the brain doesn't exert dominion over its surroundings. Instead, it is a passive conduit, receiving and embracing the ceaseless torrent of information cascading from the environment.

The Brain's Operation Predicts the Future

> *"We tend to believe that things behave in a regular manner to the extent that behavior patterns of objects seem to persist into the future, and throughout the unobserved present."* — David Hume

The foundation of survival rests upon the delicate balance of accessing sustenance and evading threats, indicating the paramount importance of information. Within this intricate tapestry, the sensory system emerges as an adept collector of this invaluable currency—information. The journey commences from stimuli, which promptly metamorphoses into electric signals traversing the neuronal network. Within this exchange, a minuscule interlude, known as the synapse,[3] introduces a slight delay in the baton-passing from sender to receiver. This choreography is not static;

---

[3]   A synaptic gap is a small gap where electric activity passes between two neurons. Each neuron can form thousands of links with other neurons, creating complex synaptic networks.

instead, it thrives on an efficiency bolstered by activity and regresses under the weight of idleness. This dance of synaptic transfer molds memory and etches the learning pathways, enabling better gathering and processing of information.

Predictability and Randomness

Amidst the ebb and flow of our cerebral currents lays predictability and randomness, influencing the ongoing brain activity that envelopes our mood and attitude. This mental disposition exerts its dominion over our sensitivity to external stimuli. Here, our elusive outlook assumes the form of a sixth sense, a harbinger that holds the power to sway or even shape our perception.

The shifting polarities of electric currents between the neocortex and the limbic system craft a symphony of cognitive inference (Deli et al., 2018). This symphony masterminds the orchestration of minimal energy configurations during transitions, ensuring the least energy conformation (Pepperell, 2018). It's this very dance that ushers the evolution of consciousness into an abstract mirror, one that faithfully reflects the environment.

Mental Topology: Within this intricate web of connectivity, synaptic topology acts as the rudder, steering incoming signals akin to the uneven terrains that guide the path of a rolling ball (as depicted in Figure 3). Frequented pathways carve valleys while the less trodden trails manifest as formidable hills (as illustrated in Figure 4.A).

Yet, this is no static landscape; alternative synaptic routes unfurl as resilient channels, offering a lifeline against minor disruptions like the

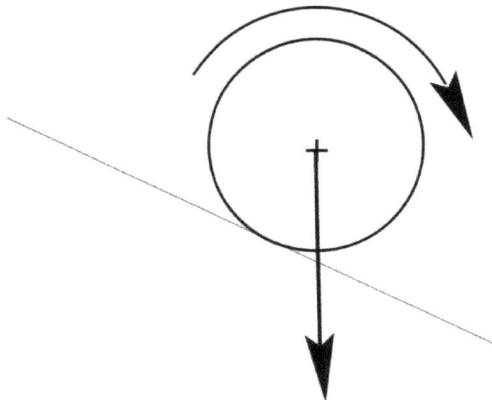

Figure 3. A ball rolling on a hillside. The ball's momentum depends on the incline.

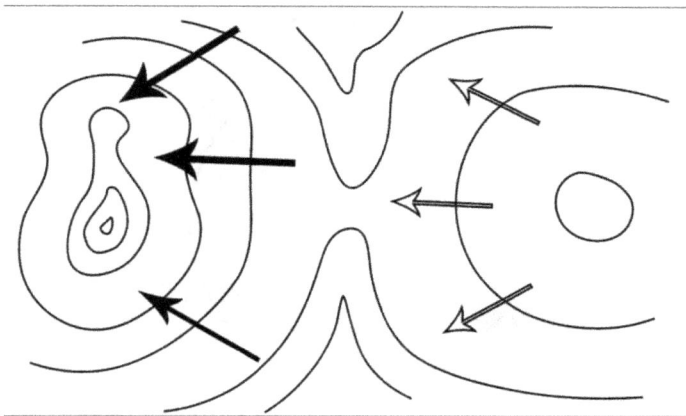

A

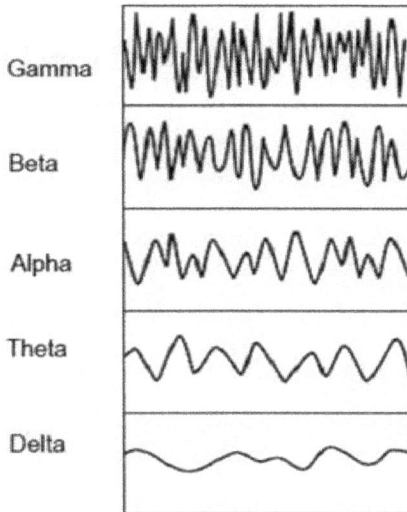

B

Figure 4. The brain frequencies. A. The distances between the lines correlate with differences in timing. Faster frequencies are closer, creating a sink (left); the contraction correlates with positive temporal curvature. Slower oscillations form a source (right) of the negative curve. White arrows indicate expansion, whereas black arrows show the direction of time pressure. B. Brain waves are classified according to their frequencies.

demise of a neuron. These secondary passages harbor the capacity for unveiling unexpected behaviors and embracing flexibility, a testament to their significance within the intricate labyrinth of complex environments. The ever-evolving contours of mental evolution are supported by psychology and social sciences.

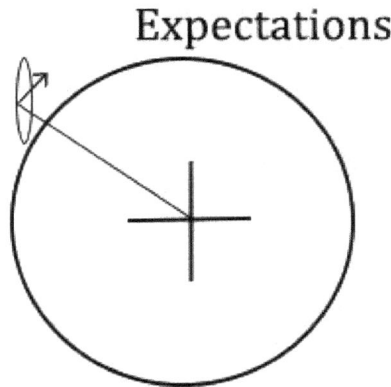

Figure 5. The mind as a gyroscope. Expectations direct motivations unconsciously. Emotions signify a deviation from expectations, and their intensity is proportional to the energy required to restore a neutral state.

Fractal Structures: In a realm of complexity, signal fragments can unveil a statistical essence of the whole. From the undulating rhythms of ocean waves to the intricate contours of landscapes, the labyrinthine formations of lung tissues, the branching intricacies of trees, and the oscillations within an electroencephalogram (EEG), a recurrent motif emerges—a symphony of self-similar patterns echoing across myriad levels, known as fractals (Zappasodi et al., 2015). These constructs are essential constituents of the physical universe and the intricate landscape of the brain (Figure 5).

From the Brain to the Mind

"Minds are simply what brains do." — Marvin Minsky

The profound inquiry into consciousness resides at the very heart of both our spiritual exploration and scientific quest to understand ourselves. Consciousness research promises to illuminate our comprehension of disorders, the nuances of decision-making, and the intricate tapestry of self-awareness. Moreover, this journey stands poised to unfurl the ethical and philosophical dimensions underpinning our societal values and frameworks.

Consciousness, the quality of the cortical brain, is a puzzling inner cosmos. Existing within precise anatomical boundaries, the brain gives rise to a mind with limitless imagination and potential. The enigmatic interplay between the mind and the brain has remained an intriguing puzzle that continues to captivate.

The quest of neuropsychology is to decode the psychological and mental phenomena as intricate products of the symphony of brain circuits. Within this paradigm, the very nature of consciousness must take on a physical guise. The mind unfurls as an ever-changing abstract reflection of the physical environment. This blueprint empowers it to harmonize and modulate behavioral responses with shifting stimuli. Within this dynamic, diverse physics frameworks can illuminate the puzzles of cognitive sciences (Déli, 2020a, b).

The cross-pollination of fields in the sciences has led to significant advances. Physics has influenced chemistry, biology, and economics by exploring matter and energy. Similarly, psychology, for more than a century, has harnessed the tools of quantum mechanics to unveil the fundamental underpinnings through which concepts intertwine, meld, and dance within human cognition (Wendt, 2015). Quantum cognition uses the mathematical formalism of quantum theory to shed light on memory and decision-making (Basieva et al., 2019). For example, the context-dependence of decisions, perception, and attitude potentially contribute to the reproducibility crisis within psychology and social sciences (Dennett, 2018).

In neuroscience, the brain can be dissected in various meaningful ways; however, the mind defies such segmentation. The mind is the most fundamental unit of intellect, indivisible by partition constraints. The realm of conscious experience is a unique privilege—a first-person sojourn into subjective perception, thoughts, and imaginings, shrouded in an enigma inaccessible to others. This exquisite process colors even our dreams, where odd sounds, scents, or lights weave themselves into our slumbering narratives. When prompted, people can smile or frown in certain phases of sleep (Türker et al., 2023).

Merely a few decades ago, the domain of the mind remained the exclusive realm of philosophical and religious contemplations. However, the advent of cutting-edge technologies affording intimate glimpses into the living brain has transfigured cognitive science into a thriving interdisciplinary pursuit, embracing the realms of linguistics, anthropology, and neuropsychology. With these tools in hand, the once elusive nature of consciousness stands poised to emerge from the shadows into the light of comprehensive understanding.

The Enigmatic Intricacies of the Human Mind

Human beings have always been intrigued by the mind's complexities. It houses consciousness, memory, emotion, and countless processes responsible for our subjective experience. The mind has bewildering attributes that, despite scientific advancements, remain perplexing. The intertwining of neuroscience with thermodynamics, quantum mechanics, and psychology presents a promising path to unraveling some of the mind's most puzzling attributes. Understanding these complexities, particularly from an energetic perspective, a sense of resting state, and the relationship between the brain and the mind, provides profound insights into our understanding of ourselves.

Energy and the Mind: The human brain is an energy-intensive organ. Considering it takes up about 20% of the body's energy despite being just 2% of its mass underscores its significance. Essential cognitive functions such as maintaining ionic gradients, synaptic transmission, and signal transduction consume this energy. Disturbances in energy metabolism can significantly affect mental and immune functions, thus making understanding the thermodynamics of the brain critical. Intriguingly, the energy demands of processes like consciousness can lead to mental fatigue, making being aware and processing information an energy-intensive endeavor.

Resting Homeostasis: While the brain is energetically active, it also has a resting state. This default mode seems to be the baseline of our cognitive existence. Without external stimuli, the mind engages in introspection, mind-wandering, and other intricate processes. This homeostatic regulation and the balance between evoked and spontaneous activations showcase the brain's adaptability. It's intriguing how the brain balances the dynamic interplay of sensory inputs with a stable, coherent resting state, ensuring a continuous experience of self-consciousness. Such a balance might be instrumental in understanding memory, learning, and the brain's energy dynamics.

Brain-Mind Relationship: The debate on the relationship between the brain (physical matter) and the mind (conscious experience) has raged for centuries. Neuroscience challenges traditional dualism but hasn't yet provided a comprehensive explanation. Recent interdisciplinary approaches, borrowing from quantum physics, have tried to capture the nuances of human cognition, showing promise in understanding the elusive nature of consciousness. This bridging of physics with psychology is not new; the influences of Jungian psychology on early particle physics are a testament to the historical intertwining of these fields.

Emotional Regulation: Another puzzling domain of the human psyche is emotion. Sigmund Freud's work hinted at how deep-rooted emotional traumas could manifest into physical and psychological issues. Emotions are central to our decision-making processes by influencing our perception of time. The interplay between thermodynamics and emotions, which could provide insights into how time perception affects our psyche, remains an area ripe for exploration.

Chapter 2

How the Brain
Produces Emotions

"Your intellect may be confused, but your emotions will never lie to you."

— Roger Ebert

Until the end of the eighteenth century, the bedrock of economies lay in manual labor. In this realm, the sinewy strength of muscles propelled tasks spanning manufacturing, agriculture, and trade. However, this era also harbored a subtle yet potent undercurrent—a space where workers harnessing their minds' power could conceive of dangerous notions capable of unsettling non-democratic governments. This intellectual potency posed a dual threat; a calculated response from oppressive regimes propagated a latent fear to subdue dissent while subjugation simultaneously created formidable barriers against the pursuit of self-realization.

The past hundred years bore witness to a profound and breathtaking transformation. This technological upheaval supplanted the sweat-soaked toil of physical labor with the prowess of machines. Simultaneously, the realm of research illuminated intriguing revelations about the pivotal role of attitude and mood in performance. Similar to a burdensome tax, the toll of conflicts not only saps morale but also corrodes work efficacy. These revelations resonated within a swiftly evolving professional landscape that demanded intellectual acrobatics and the cultivation of creativity from workers. Companies attuned to this dynamic safeguard their core values by weighing the human element[4] and cultural resonance in their hiring decisions.

---

[4] The human element is the developer and custodian of the organization's culture, values, and mission.

Nonetheless, for all their strides, societies remained planted within the embrace of rigidity. As information technology (IT) professionals envision sweeping societal transformations, the bulk of society clings to the familiarity of their time-honored ways, a complex juxtaposition that engendered profound social inertia in the face of change. The problem is severe and requires a complete transformation of social dynamics.

This intricately woven narrative leads us to an imperative query—how can families and schools recalibrate their motivational tools, transitioning from the dated "carrot and stick" paradigm to the nurturing embrace of respect and love? In this journey, our worldview—an intricate mosaic etched over millennia through the indelible marks of wars, oppression, and the caprices of natural disasters—defies the prospect of swift metamorphosis from a realm of fear to one of trust. A striking example surfaces within school curricula—meticulously tailored to mirror technological progress, yet often bereft of the essential tools to impart the delicate nuances of social intelligence.

As a result, the mantle of social learning—a torchbearer transmitting a society's cultural legacy to the next generation—must transpire within the home. Within this familial crucible, orchestrated dialogues of verbal and non-verbal communication, interwoven with behavioral modeling, serve as the conduit for the passage of cultural wisdom from parent to progeny. Yet, within this intricate choreography lies an inherent dichotomy. In this space, children, like little sponges, absorb not just the nurturing aspects but also unwittingly assimilate fears, prejudices, and corrosive narratives. Parents can inadvertently undermine their children's self-assuredness and human potential in this dance of influence. Understanding how we can transform the dynamics of the home commences with the contemplation of the nature of emotions and their intricate genesis.

The Nature of Emotions: Nature's canvas unfurls, driven by the entropic[5] currents orchestrating grand-scale organization. Within the realm of neural dynamics, disturbances trigger a symphony of resting recovery to unfold through compensatory emotional responses. These responses form a delicate dance—hunger cues the instinct for nourishment, a fall propels the quest for rebalancing, and a threat sparks the search for sanctuary. Just as the gravitational forces maintain celestial equilibrium, the yearning for psychological security is intricately entwined with the fabric of emotional balance. Within this symphony, emotions rise as the intimate custodians of personal, subjective homeostatic regulation.

Like the equilibrium position for the pendulum, the brain's resting state represents a mental balance (Zanin et al., 2020). Embedded within

---

[5] Entropic directionality is the physical environment's statistical tendency is to move toward equilibrium.

the involuntary nature of emotions is a profound revelation—an essence that transcends mere feeling and takes on the aura of energy (Kao et al., 2015). Furthermore, emotions represent a kaleidoscope of multi-dimensional representations and varied cultural and brain activity profiles (Nazmi et al., 2020). Nevertheless, emotions' essence can be distilled into positive or negative energy states (Kao et al., 2015). This polarity of emotional valence dances between positivity and negativity (Hesp et al., 2021), representing an instant feeling for or against (Surov, 2022).

The Fundamental Nature of Emotions

"If we resist our passion, it is more due to their weakness than our strength." — François de La Rochefoucauld

In 1928, Phillip Bard surgically removed cats' cortical hemispheres. What followed was a revelation that morphed the creatures into vessels of unrestrained fury, responding with disproportionate anger to even the faintest stimulus. This phenomenon, famously dubbed "sham rage", unfurled as a potent testimony to the indispensability of context—woven intricately by the associative realms of the frontal cortex—in shaping the contours of an appropriate emotional response.

So, how do emotions form? While the realm of positive feelings often feels like magic, the reality is that we're not immune to the shades of blue, the sparks of aggravation, or the tears that well up in response to tragedy. While the tapestry of emotional expression takes its cues from the interplay of cultural, religious, personal, and social factors, our bodies serve as unwitting informants. The canvas of our physiological reactions—the shift in skin color, the cascade of tears, or the cadence of laughter—betray our internal state. The sinking sensation that grips us might whisper of humiliation or the stark isolation we grapple with, while taut muscles and a racing heart might sound the siren of imminent peril.

Nevertheless, emotions are a universal undercurrent connecting emotional animals (Figure 6). The capacity to produce feelings is intertwined with the privilege of recognizing these very feelings. Aesop's ageless parables mirror this essence—projecting the quintessence of pride, compassion, and shrewdness onto the animal realm. Concern for others rests upon the potent foundations of "feeling the pain of others", an empathy that resonates across species, uniting us in the universality of emotional experience.

Emotions persist as a steadfast yet not always conscious state of mind, a cornerstone upon which our psychological equilibrium, wellness, and health hinge. Insight into this realm can reverberate across domains, enriching the tapestry of our relationships, rendering the learning

Wong-Baker FACES® Pain Rating Scale

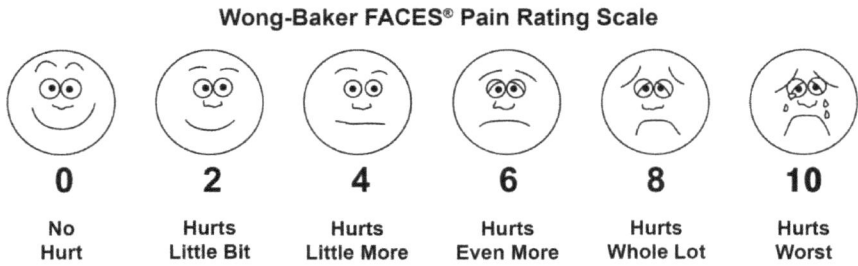

0	2	4	6	8	10
No Hurt	Hurts Little Bit	Hurts Little More	Hurts Even More	Hurts Whole Lot	Hurts Worst

©1983 Wong-Baker FACES Foundation. www.WongBakerFACES.org
Used with permission. Originally published in *Whaley & Wong's Nursing Care of Infants and Children.* ©Elsevier Inc.

Figure 6. Wong Baker pain scale. This pain scale was initially developed for children to rate their pain. Wong-Baker FACES® Pain Rating Scale, with permission from http://www. WongBakerFACES.org.

environment more fertile, amplifying professional prospects, and infusing happiness into our homes. This essence serves as a catalyst for creativity, an indispensable wellspring in the bedrock of our innovative society.

Mental Energy

A perpetual influx of sugar and oxygen sustains the intricate neural network, serving as the lifeblood that fuels its operations. However, the scale of human achievements remains a diverse panorama, a testament to the profound impact of motivation—an elusive trait that demands unwavering persistence. Within courage and determination, mental energy takes center stage, illuminating individual disparities in enthusiasm, self-assurance, learning, and creativity (Di Domenico and Ryan, 2017). Mental energy has an enduring quality intertwined with intrinsic motivation, a potent harbinger that forecasts performance trajectories.

Mental energy is related to emotional intelligence—an attribute intricately woven through the tapestry of individuality. This dynamic essence transcends the confines of a singular moment, extending its influence across the spectrum of personality development and well-being landscape (Ryan and Deci, 2017). Notably, the ebbs and flows of mental energy, akin to the tides that shape oceanic landscapes, entail a thermodynamic expenditure—a testament to the intricate interplay between energy dynamics and the nuanced realm of cognition.

Like a boat navigating the vast expanse of an ocean, maneuvering the myriad currents of society necessitates confidence. Much like a sturdy vessel battling the capricious waves, the journey of intelligence finds its cornerstone in a steadfast mental equilibrium. As tumultuous seas can spell the doom of a diminutive craft, intelligence relies upon a secure and unwavering balance of the mind.

Mental energy is a long-term ability related to intrinsic motivation, which predicts performance and plays a vital role in personality development and wellness across the lifespan (Ryan and Deci, 2017). Significantly, mental energy is the brain's structural quality; therefore, mental energy changes are associated with a thermodynamic cost.

The Role of Emotions in Behavior

"What drains your spirit drains your body. What fuels your spirit fuels your body." — Caroline Myss

A scattering experiment propels particles or radiation, such as light or sound, toward a medium, redirecting the particles, absorbing them, or permitting them to journey unscathed. Measuring the particles' outgoing pattern unveils a trove of insights about the inner structure and properties of the medium. In a parallel realm, animals engage with the sensory stimuli akin to the incoming radiation in the scattering experiment. Primitive animals react predictably, offering a window into their simple organizational constructs.

Contrasting this simplicity, the stage of emotional reaction unfurls a tableau that reaches deep into the annals of history—the history of one's journey, the reservoir of mental energy amassed, and the fleeting terrain of the psyche's present state. Like the delicate interplay of particles in the scattering experiment, emotional responses mirror the inner landscape, echoing the tides of time, the energy ebbs and flows, and the poignant essence of the immediate moment.

Just as physical imbalance triggers a flurry of responses, emotional and mental instability sets off a cascade that manifests as anxiety. This emotion ignites various regulatory frameworks within us. Intricately intertwined with our belief systems, these frameworks stand sentinel, striving to preserve our psychological equilibrium. Within this realm, emotions emerge as forces of motivation, surging through every facet of our lives, infusing each stride, choice, and endeavor with energy. Nevertheless, this unstoppable power unfurls its influence in the most nuanced ways.

The Role of Feelings in Thoughts, Performance, and Intellect

In classic models of thinking and decision-making, feelings often stand relegated to the domain of irrationality. However, the tides of the past decades have ushered in a paradigm shift within neuroscience, uncovering the powerful and pervasive role of emotions in cognition, such as decision-making. Emotions are an inherent part of the self, mirroring every shift and every alteration of our environment, an essence as innate to our psyche as the very breath we draw.

- Our intuitive gut feelings frequently emerge as the architects of wiser decisions than the calculated constructs of rational reasoning. These emotions are a continuous presence that wields its sway across the myriad facets of life—guiding our culinary choices, determining our sitting positions, and steering our needs, desires, and laments. Within the symphony of conscious existence, no minute, no second, is devoid of the undulating currents of emotion.

- The study of emotions uncovered the role of relationships at the foundation of our psyche. The energy we invest in forging bonds, whether with people or tangible elements, makes losses painful. As physical pain sets our body's limitations, spiritual distress casts its shadow upon our intellectual freedom. A pursuit to evade the specter of pain can sometimes spiral into paranoid self-avoidance.

- Emotions are intimately intertwined with our core environment. Supportive, loving experiences weave a cloak of trusting embrace and safety. Within the folds of love, faith, and pride, emotions become the fuel that kindles the flame of intellect, breathing life into our aspirations and endeavors.

- Across the tableau of history, luminaries of various domains— teachers, politicians, artists, and even the shrewd maneuverings of con artists—have harnessed the latent power of emotions to manipulate people, even entire nations. Within this intricate dance, a poignant truth surfaces—people and animals, as the echoes of emotional energy reflect back the essence they receive.

- Nestled as the singular driving force of mammals and birds, emotions provoke a profound sense of permanence. Within their tender embrace lies the potency to shape destinies, the ability to kindle greatness, and, at the darkest fringes, their crushing weight can project the desire for suicide. As such, emotions lay the bedrock of intellect—a realm that, by its untamable essence, necessitates not control but mastery.

- With the lens of evolution, emotions emerge as an essential tool—an intricate facet of the brain's sophisticated energy network, the linchpin of survival. Emotions are survival tools, crafting the foundation upon which intellect flourishes. This bridge integrates the symphony of the mind with the ever-shifting landscapes of the environment.

Automatic Behavior

"Chains of habit are too light to be felt until they are too heavy to be broken." — Warren Buffett

The neural system is at the helm of the intricate symphony of intuitive and conscious behavior. Within this elaborate dance, we discern a spectrum ranging from swift reflexes to contemplative decisions, each echoing the increasing depth of conscious and emotional engagement.

1. Autonomic Perceptions

Even amidst the shroud of night, the mind remains attuned to the ceaseless inflow of environmental input. A symphony of physiological reactions and subtle hormonal shifts plays out beneath the veil of awareness—expressions etched upon our faces and the tautness of our muscles tell tales of our true feelings, often contrary to our conscious intentions. In the silent corridors of the mind, experiences are etched anew through the strengthening of connections among specific neurons, forging transmission pathways. This learning unfolds as a spontaneous process, weaving itself without conscious intervention. A familiar sight—a red hexagon by the roadside—instigates the automatic identification of the stop sign, an emblem of such learning.

2. Automatic Actions

Learning to ride a bike or drive a car formulates the neurons' precise activation sequence. At the level of competence, the reins of emotionally driven conscious control loosen, ceding ground to spontaneous action, unfurling beyond the reach of deliberate intent. Whether in a debate or on the tennis court, performance hinges upon near-instant, millisecond reactions to the demands of the environment. Astonishingly, these reflexive actions, albeit wielding considerable power over our lives, can diverge from the paths laid by logic, reason, and even conscious intention.

In our dynamic existence, moments often rush past without affording the luxury of deliberate analysis or contemplative thought. These flashes of intuitive decisions, akin to minimal-energy formations, serve as cognitive shortcuts, preserving energy and time and bequeathing mental vigor. Language generation, for instance, is a symphony of automatic phrasings, freeing consciousness to thread meaning, social context, and emotional nuances—resulting in variances when addressing a boss, a parent, or a friend.

Automatic actions do not involve feelings and do not generate a new memory. Like water flowing through an open pipe, autonomous actions are the synaptic freeways of the brain. Once mastered, even complex actions, such as flawless star performances, can arise with spontaneous ease. Creative shortcuts give rise to the master's assured brushstroke or the pianist's precise finger movements. Thus, artists do a poor job of explaining their work.

3. Habits

Habits are sculpted through the crucible of reinforcement and repetition. Within these finely-honed muscle movements resides a reservoir of intuitive precision, bestowing us with confidence within the familiar realms of behavior. They weave the fabric of our morning rituals, define our customary environments, and shape our social etiquette. Yet, an interruption in the fluidity of this automatic signal flow can lead to unraveling the very action sequence—spawning errors and sowing seeds of confusion. A crying child or an unexpected phone call can taint the sacred dance of the morning coffee routine.

Routines also include rigid, formal, and repetitive rituals endowed with symbolism and meaning. These can improve individual performance and group cohesion.

The struggle to break free from the clutches of smoking or other addictions speaks volumes of habits' formidable dominion over conscious intention. Bad habits, resilient and tenacious, are arduous to relinquish; conversely, cultivating virtuous habits requires persistent conscious effort. Herein, the tapestry of one's routines knits the very texture of life, a testament to their power in shaping the contours of existence.

The Modern, Cohesive View of Consciousness

> *"I... a universe of atoms, an atom in the universe."*
> — Richard P. Feynman

Self-perception, that intricate tapestry of our being, finds its roots in the interplay between the senses and the inner workings of our organs. The softness of sand beneath our feet, the cold kiss of raindrops—these sensations meld into a cohesive reality. As Thomas Nagel so aptly phrases it, the elusive "what it is like to be" forms the heart of qualia, an enigmatic realm of highly subjective mental landscapes that constitute the self.

Mental Unity

Amidst the cacophony of incoming information, a singular conscious experience emerges—a phenomenon that has dominated philosophical discourse since the days of Descartes and Kant. This unity arises from the twin threads of attention, capable of embracing only one facet at a time and the tendency of even tumultuous mental content to coalesce into an intrinsic oneness. The most disparate, bewildering, or disorderly information weaves into a unified perception tapestry (Tring et al., 2023). While mind-altering substances, neural injuries, or the march of neurodegeneration can fray the threads of coherence, the self-identity of the psyche remains.

In the case of split-brain patients, the severing of connections between cortical hemispheres births a dual existence, each half housing independent perceptions and a distinct sense of self. These halves harbor separate experiences, intentions, and preferences. However, the splintering of the brain further dismantles consciousness. In instances of stroke, the demise of neurons can obliterate significant portions of the brain. Yet, as untouched areas assume the mantle of the defunct, the shrunken neural volume orchestrates near-complete or full functionality.

Over time the Self manifests remarkable unity. The trajectory of the self undergoes dramatic metamorphoses from infancy to the embrace of old age, yet the core of self-identification remains constant. This sense of unity evokes spiritual awe for the cosmic oneness of being.

Beliefs: Beliefs stand as unwavering pillars, the bedrock upon which our mental universe is built. An assault on these personal bastions triggers profound emotional reactions. The human need for cognitive consonance, as Krause (1972) describes it, is an irresistible force; no amount of evidence can dislodge these convictions. Facts wield no balm against hopelessness, nor do predicaments shatter the glass of optimism. Without belief, we surrender, and setbacks transmute into irreversible defeats. Moreover, when our convictions diverge, they can kindle intense personal and societal strife.

Festinger's Cognitive Dissonance Theory: Festinger's cognitive dissonance theory (1957) postulates that incongruence between two beliefs begets cognitive conflict, termed dissonance. This dissonance, the disjuncture between conviction and behavior, can birth madness or abandonment of our core values, hinting at the trajectory toward criminal pathology.

The Holographic Mind

Through an intricate interference process, Holography manages to preserve the depth of a three-dimensional scene. In a parallel manner, our sensory perception captures the profound complexity and depth of the physical world, all neatly encoded upon the canvas of the cortical surface. When it comes to decision-making, we merely skim the surface of our unfathomably intricate subconscious. Much like a hologram's ever-shifting perspective, the meaning we attribute to things is intrinsically tied to the web of experiences we've woven. A seemingly simple word like 'ball' can conjure a cascade of associations—baseball, ballpoint pens, paintball, or the novel combinations that human creativity constantly concocts.

Consciousness, acting as a distinct observer, maintains its separate status over the self-regulating mechanisms of the brain.

Now, let's delve into the intricacies of the brain's mechanisms that underpin our versatile psychology. Why does the governance of our thoughts often feel like grasping at elusive shadows? Why do our beliefs seem to elude the reins of conscious control? The answers to these profound inquiries about consciousness find their roots in the self-regulatory essence of the human mind. The mind operates as a homeostatic organ at its core, ensuring equilibrium and stability amidst the dynamic currents of experience. From this bedrock, a solid foundation for consciousness science, built upon rigorous physical principles, must naturally emerge.

The Resting State

"We want the facts to fit the preconceptions. When they don't it is easier to ignore the facts than to change the preconceptions."
— Jessamyn West

Hans Selye (1970)[6] pioneered stress research by closely examining the universal reactions of patients to illness. His groundbreaking studies revealed physiological shifts in response to prolonged stress. It's becoming increasingly evident, through findings in psychology and social sciences, that psychological evolution mirrors the ever-shifting environment we inhabit.

In many ways, we navigate life like finely tuned antennas, attuned to the frequencies of our expectations. Amidst the dance of our surroundings, the mind wields a remarkable self-regulatory power, ensuring a steady, self-sustaining equilibrium. The brain's spontaneous fluctuations in the absence of stimuli form a restful baseline, akin to a car idling.

Mental Homeostasis: Human beings possess a unique knack for preserving psychological equilibrium. This ability, operating automatically and independent of conscious awareness, enables the mind to counterbalance disturbances imposed by the external world. It forms the bedrock of the mind's homeostatic regulation (Smitha et al., 2017). As past states fade and the future remains uncertain, consciousness—representing the transient present—changes over time.

---

[6]   Hans Selye introduced and popularized the idea of stress.

This self-aware readiness stems from our identification with the body through sensory perception. Even as sensory input disrupts subsequent layers of regulation within the brain, the mechanisms of self-regulation work in harmony to maintain the constancy of the resting state's personalized perspective (Wolff et al., 2019). This is achieved through precise regulatory integration (Zanin et al., 2020). Unruly thought processes converge to craft a coherent self-narrative (Smitha et al., 2017), constructing a subjective, transcendental, and uniquely personal first-person experience (Kolvoort et al., 2020).

The brain adeptly translates spatial signals from the sensory system into the temporal cadence, where the slow oscillations stand as steadfast pillars, with precise timings dictating the rhythm of activation. The resting-state network effectively integrates information into abstract representational hierarchies, spanning a vast geodesic distance from primary sensory and motor regions. When stimuli are absent, this fertile ground becomes a realm of mind-wandering, autobiographical memory, future projections, and introspection. This abstract world model is the foundation of the mind's self-consciousness and boundless imagination (Tring et al., 2023; Wolff et al., 2021).

Fundamentally, survival—hinging on nourishment and security—is intricately tied to sensory input. Contributions to evolutionary fitness are often the primary argument for emotions. However, a novel perspective emerges when viewing the brain as a self-regulating system governed by physical laws: emotions represent energy imbalances that steer behavior toward restoring equilibrium. At the same time, they drive changes within the brain's neuronal landscape. The energy expenditure needed for updating synaptic organization transforms sensory processing into a thermodynamic endeavor.

The Resting State as a Physical Field

In physics, the field[7] is a surface with defined energy at every point of space and time. The field's integrity hinges on a seamlessly interconnected map, and any abrupt discontinuity prompts immediate correction. In the social sphere, fields manifest as intricate patterns of cultural norms, national identities, age-based expectations, and class privileges.

Place cells, which encode physical location also assign social relationships (Forli and Yartsev, 2023). These societal overlays are imprinted upon our minds' resting state, contributing to our subjective beliefs' formation. Interactions are set into motion when these internal

---

[7] A field represents an energy value with varying strength from point to point. They resist direct experience and can be assessed only indirectly. For example, we cannot see gravity but feel its effect.

beliefs deviate from external expectations. These interactions serve as catalysts for updating our personal belief map, thereby instigating a continuous shift in the landscape of our mental states. Simultaneously, the social field itself undergoes revision through the collective experiences of its members.

The resting state, acting as a personalized projection of the encompassing social field, encapsulates our perceived behavioral expectations. In essence, it is a guiding force that shapes our social behaviors. Just as particles' interconnected energy dance governs the curvature of the gravity field, the resting state wields a similar power in governing social conduct.

Mind-wandering, the Path of Insight

"Your worst enemy cannot harm you as much as your own unguarded thoughts." — Buddha

Over the past few decades, rapid advancements in neuroscience have unveiled a profound correlation between neural activities and psychological processes.

It has become evident that focused and negative mental states carry a substantial amount of information, rendering them energetically more demanding compared to positive and neutral conditions (Trevisiol, 2017). The accumulation of stress (Xu et al., 2023) raises what can be referred to as emotional temperature[8]—a social analog of temperature. This phenomenon parallels how negative emotions are born from the same pain receptors, whether triggered by a minor physical discomfort like bumping an elbow, a material loss like a wallet, or the loss of a loved one. This shared origin in pain receptors gives rise to tightening and pressure (Figure 5), a precursor to deterministic[9] behavioral responses (Sakai, 2020) that can ultimately lead to cognitive fatigue.

The progression of mental weariness eventually transitions into what is commonly called "mind-wandering.[10]" This mental wandering also occurs during moments of boredom, such as difficulty falling asleep at night, monotonous conferences, or mundane work tasks. Such mental disengagement experienced while driving or at work can

---

[8] Temperature is the particles' kinetic energy and social temperature is the need for competition (Déli, 2020a). Supply abundance allows cooperation forming low social temperature (Tkadlec et al., 2020), but lack causes competition, deterministic high social temperature (Stewart and Plotkin, 2012).

[9] In a deterministic system, each state evolves with a high probability to a subsequent state (Varley et al., 2021).

[10] Mind-wandering or daydreaming is transient, unfocused mental time travel emotionally connecting seemingly unrelated but personally relevant events past and future.

have dire consequences, including accidents and errors with significant repercussions. This prompts a critical question: Why does the brain engage in seemingly wasteful and perilous activities?

A Path to Insight

Mind-wandering encompasses more than simply shifting away from focused attention. Rather than being a mere time-wasting activity, it can be likened to gathering rain clouds, which eventually turn into rain. Just as the qualities of the terrain and the wind influence rainfall, mind-wandering filters new information through our existing beliefs, dreams, visions of the future, or reflections on the past. These mental wanderings continually update our resting state, maintaining our mental stability and forming insight into our social position (Smitha et al., 2017). This process of piecing together coherent narratives from fragmented, disjointed elements forms the basis of our gut feelings—the initial emotional filter through which we perceive stimuli.

The pessimistic mind is stuck in the past. Pessimism is a broken record that ensnares the mind in repetitive and self-defeating thoughts. In contrast, a flexible and agile brain engages in activities that enhance performance by generating new associations, ideas, and discoveries. Consider the purposeful and enriching mind wanderings of creative geniuses like Mozart and Einstein.

Using Stimulants: Experimentation with brain stimulants has been a part of the human experience for millennia. In various cultures, substances like coffee, tea, and alcohol have become sanctioned components of social gatherings. Psychedelics and hallucinogens, on the other hand, can profoundly alter perception, mood, and cognitive processes, leading to distortions in thinking, sense of time, and emotional experiences. They are being studied for their potential to address various medical issues, including pain management, addiction treatment, depression therapy, and the induction of brain plasticity. However, their addictive nature and long-term effects on executive functions such as memory, learning, and motor and verbal abilities urge caution (Carr and Schatman, 2019).

The Evoked Cycle

> *"Experience is food for the brain."* — Bill Waterson

The evoked cycle is the foundation for the brain's interaction[11] with its environment. The energy expenditure associated with perception firmly

---

[11] Interaction modifies neuronal connections in the brain, which correlates with a change in cognition or memory.

situates the brain within the energy dynamics of its surroundings. Consider the act of seeing a road sign (activation), which prompts you to turn the wheel (response). Similarly, the sensation of hunger (activation) encourages the preparation of a meal (response). These evoked activities ripple across the cortical surface like waves on a pond, leading to modifications in the synaptic connection map and a continual update of the resting state.

Neural activation can be likened to pulling a spring, drawing a bowstring, or setting a pendulum in motion. Just as potential energy drives the restoring force in a pendulum, bow, or spring in the brain, minute potentials facilitate the movement of energy (or information) between the cortex and the limbic brain. This energy-transfer functions to restore the resting equilibrium, thereby preserving the fundamental constancy of the mind (Smitha et al., 2017).

Similarly to a ball's ability to move in any direction on a flat surface, resting thoughts possess a remarkable degrees of freedom.[12] In physics, "degrees of freedom" pertain to the number of independent variables governing positions or motions in space. However, much like a ball tracing a predictable trajectory on a hillside (Figure 4), sensory inputs from the eyes, ears, tongue, and skin follow well-defined paths toward the associative regions of the brain. This leaves little time to consider alternative options, and conversations or behavioral actions often occur instinctively. Our attitudes, rather than conscious intentions, guide our spontaneous actions in these situations.

Entropy

"A brain is only capable of what it could conceive, and it couldn't conceive what it hasn't experienced." — Graham Greene

An agent's capacity to generate work is intricately tied to its internal organization. In systems isolated from external influences, the constituents naturally evolve toward a stationary state where no observable changes or work occur. This equilibrium state is referred to as high entropy.[13]

The concept of entropy was introduced by Rudolf Clausius as a form of energy incapable of producing work. This notion was redefined in statistical mechanics as the system's microstates (Martyushev, 2013), signifying the various possible configurations linked to thermal disorder (Kostic, 2014). In the 1940s, Shannon extended the concept of entropy to information science, defining it as a measure of surprise potential (Shannon, 1948). More recently, investigations have explored the

---

[12] In cognition, mental freedom refers to a person's possible behavior choices.
[13] Entropy is the tendency to disperse energy or particles in a system.

significance of information and entropy in biological systems, including the living brain (Koonin, 2016).

In our daily lives, entropy is responsible for cooling our morning coffee, the aging of vehicles, and the corrosion of equipment. Throughout human history, the struggle against entropy has been a persistent challenge.

Entropy in the Brain

The consideration of entropy within the context of the neuronal system has provided fresh insights into understanding brain functionality. The coherent low-frequency resting oscillations of the default mode network (DMN) provide a wealth of degrees of freedom (Smallwood et al., 2021). These degrees of freedom pertain to the number of accessible neural states and have predictive value for complex behavioral performance, intellectual capacity (Yang et al., 2019), and creativity (Shi et al., 2019).

Unlike its definition in the physical world, entropy in the brain aligns with mental energy, serving as the source of intellect. While entropy may signify disorder in your room, it fosters cognitive flexibility within your brain.

Living systems rely on metabolic networks to counteract erosion and decay. The evolution of the cortex introduced even subtler energy regulation, maintaining the high entropy[14] resting state.

The activation paths' high degree of freedom makes resting thoughts transient and hard to retrace. Simultaneously, their spontaneous nature gives rise to innovative associations and sophisticated behaviors. It takes energy to rein into focus these fleeting associations. Hence, the interplay between entropy and energy profoundly influences actions and decisions, a topic explored further in the subsequent section on emotional regulation.

Emotional Regulation

A hundred years ago, Sigmund Freud's systematic investigations paved the way for scientific psyche analysis. Psychoanalysis delves into the concealed structures of personality, illustrating how emotional traumas can lead to motivational, psychological, and physical issues. Although we take pride in our consciousness, the short-lived and fleeting cognitive focus relies heavily on emotions to color virtually every thought, judgment, and decision. Even when feelings remain beneath the threshold of consciousness, they impact what we perceive, hear, and see (Xu and Schwarz, 2017), catalyzing creative brilliance, driving criminal acts, or triggering self-destructive tendencies like suicide (Beall and Tracy, 2017).

---

[14] Mental entropy correlates with the lack of predictability and a high degree of freedom.

Warm-blooded animals, including humans, rely on emotions, vital in parental care behaviors (Farmer, 2020). Interconnected regulatory pathways establish a strong correlation between emotion and thermoregulation, evident in responses like shivering under stress or perspiring due to fear (Grigg et al., 2021), highlighting the central regulatory role of emotions.

The Energy Characterization of Emotions

"Most every image in the main procession we call mind, from the moment the item enters a mental spotlight of attention until it leaves, has a feeling by its side." — Antonio Damasio

Sham rage has underscored that emotions are a perpetual presence within the mind, even if not always consciously acknowledged. While emotions align with every shift in the environment, their intelligent response necessitates the cortical interpretation of the stimulus (Figure 7). Emotional valence represents the energy of the stimulus, and the context provides its meaning. Thus, emotions can be categorized based on their energy demands. Negative and focused states require higher energy than positive and neutral emotions (Kao et al., 2015).

Comparable to canaries in coal mines, emotions act as indicators of unstable energy conditions, urging adaptive changes to restore equilibrium. Therefore, emotions exert a commanding influence over our behavior, often operating below conscious awareness to influence our perceptions, auditory experiences, and thoughts. Hence, emotions effectively signify shifts in the brain's energy or synaptic dynamics.

Positive feelings broaden our comprehension, radiating enthusiasm and generosity. Signal sparsity expands into negative curvature, increasing cognitive freedom (Figure 5) and potential (Fry, 2017). However, like delicate soap bubbles, their expansive essence makes them ephemeral. A smile, an awe, or a surprise fades swiftly after its initial appearance.

Negative emotions are information-rich (Kao et al., 2015), and their synaptic connections serve as attractive loops, perpetuating deterministic, repetitive thought patterns. These patterns manipulate our psychological landscape with their partial viewpoints. Additionally, emotions like shame, remorse, regret, and hate can linger within the synaptic network for years. Unexpectedly, their energy can gather momentum, transforming into anger, madness, and destruction.

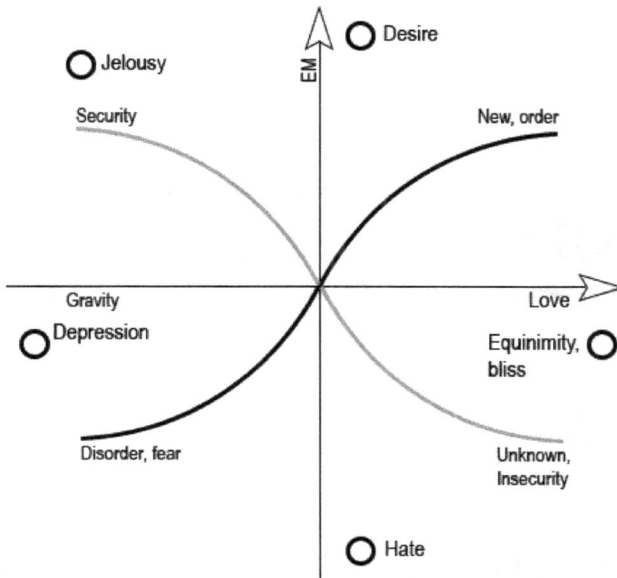

Figure 7. Energy values of some emotions. The two-dimensional representation of emotional magnetism and temporal gravity. The horizontal line represents temporal gravity, which varies between pressure on the left and expansion on the right, while the vertical coordinate represents emotional electromagnetism, which changes between desire at the top and aversion at the bottom. Just like in a computer game, where a plane disappears on the left and reappears on the right, the two sine curves that cross at zero can flip at their extremities - from security to curiosity and to fear. In an instant, a fight can transform into a powerless flight. However, love has the power to overcome fear and inspire a sense of security.

While conscious reasoning may cloak genuine motivations, compromised energy regulation can detrimentally impact personal, professional, relationships, and psychological and physical well-being. Mental degradation can contribute to hormonal imbalances, immune issues, and mental disorders.

The ensuing sections will delve into emotions' classification and operational intricacies, encompassing their motivational generation and the reasons behind our recurrent struggles to execute our intentions.

Chapter 3

The Mind as a Particle

"The universe gives birth to consciousness, and consciousness gives meaning to the universe."

— John Archibald Wheeler

The wonder we feel staring at the night sky is a testament to our desire to understand our place in the universe. Although neuroscience can probe the human brain in detail, unraveling the intricate web of neurons, hormones, and cognitive processes that underpin our conscious experience, it is at a loss to explain consciousness.

Our ability to process and adapt to new information underscores the flexibility and dynamic of the mind. Through this adaptability, we can intuit the laws of physics and translate them into intelligent behavior, technological innovations, and philosophical insights. In essence, our journey from gazing at the stars to pushing the boundaries of knowledge exemplifies the power and potential of the human mind.

Perception as a Path to Understanding: The natural world has fascinating examples of adaptability and intelligence that may not align with our traditional definitions. Different species evolve unique strategies to interact with their environment. With its limited vision, the spider relies on detecting vibrations in their webs to identify captured prey. The comparison of a spider's web-centric universe to our own human cognition and consciousness is a compelling one.

The lack of scientific consensus on consciousness highlights the need for a radical approach. It's impressive how a lowly spider can process and interpret external stimuli with such efficacy and intricacy. Processing sensory information can construct a comprehensive model of one's social and physical surroundings. The experience of the physical world leads to the seamless integration with the environment. Finally, adopting the physical laws as the operational principles of perception engenders striking similarities between consciousness, material systems, and the universe.

The Mind Regulates Its Environment

The Good Regulator Theorem, initially proposed by Conant and Ashby in 1970, asserts that for any system to regulate itself efficiently, it must possess a model that mimics its operations. In the case of the brain, experiences build mental complexity as an environmental model. Intelligent modeling requires environmental insights beyond superficial projections. The continuous engineering of superior regulation requires the brain's innate intuition of the physical laws. Recent neuroscience studies have highlighted striking similarities between the organization of the physical world and the activities of the brain, as noted by Deli (2020a, b).

Libet's early experiments on free will showed a close correlation between volition and the brain's background electrical fluctuations (1985), revealing the relationship between motivation and energy regulation in the brain. As such, thermodynamic considerations can explain the nature and origin of emotions and intellect.

The brain, a veracious energy-consuming organ, integrates into the environment's thermodynamic cycle via the sensory and muscle systems. Therefore, it can be analyzed based on information and energy exchange, much like any other physical system, such as an engine. The question that arises is how far do the similarities between the brain and the physical world extend? The book, *The Science of Consciousness* (2015), argues that material interaction increases cosmic complexity through physical, chemical, and biological evolution, ultimately leading to the emergence of an intelligent mind. Therefore, consciousness is the child of the universe, inheriting its structural and operational organization.

The Surprising and Unexpected Organizational Similarity of the Mind and the Universe

"The universe is built on a plan the profound symmetry of which is somehow present in the inner structure of our intellect." — Paul Valery

The emergence of life, particularly intelligence, in the material universe is a question that has been a source of meaning and wonder for human existence. The concept of the soul, as provided by religion, was an initial answer to this question. As philosophers delved into this issue, they formed divergent, unresolved viewpoints. One such example is dualism—the intuitive belief that conscious experiences are separate from the material world.

More recently, the discourse has been dominated by scientific perspectives, acknowledging the physical nature of the mind. While the body occupies physical space, searching for the mind within the brain or any tangible location proves futile. Even though self-perception is linked to the body and can be influenced by social and virtual cues, the mind lacks attributes of physical dimensions such as length, weight, or height.

Characterizing the mind functionally proves elusive. Our mental discourse remains beyond our complete control, giving rise to spontaneous and involuntary thoughts, internal dialogue, emotions, and actions. Errors, irrational choices, and even foolish decisions result in feelings of regret, shame, and remorse. From the most abstract theoretical ponderings to the persistent recollections of pain or humiliation, consciousness merely represents the visible surface of an extensive mental reservoir. The mind is a self-regulating independent organization resistant to traditional scientific investigation. Still, poetry over millennia provided mesmerizing glimpses of the depths beneath this impenetrable mental exterior.

A hundred years ago, the intricate beauty of Ramón y Cajal's drawings brought brain organization into focus. The Spanish neuroscientist thought that the cosmos' awe-inspiring complexity is a mere reflection of the brain. The truth is even more remarkable. Life's reliance on the physical world is a fundamental fact of biology. Gravitational pressure lends stability and balance, ensuring cell growth and organizational symmetry. The explanation of the mind's mysterious, yet unequivocally physical nature, might lie in particle physics.

The Mental Particle

Imagine the universe as a grand, intricate puzzle with pieces so small that they're beyond our everyday perception. The tiny building blocks of this puzzle are matter particles called fermions.

Fermions are half spin[15] constituent particles of matter. Half spin is the wave function's momentum (Ohanian, 1986), pointing in the direction or opposed to a magnetic field (–1/2). The conservation of momentum requires that only opposite spin fermions (up and down spin) can occupy the same quantum state, as expressed by the Pauli exclusion principle. Although the Pauli exclusion principle defines a unique energy level to each electron, all electrons have precisely the same unchanging properties of mass, charge, and spin.

This principle necessitates the atom's different energy levels, making it possible to form bonds in molecules. Adopting these spin attributes as psychological spin (Deli, 2023), leads to our inclination for interpersonal comparisons, competition, and establishing hierarchical social structures. Hence, the mind's social qualities can emerge from its fermionic essence.

---

[15] A half spin particle resembles a tiny bar magnet.

Fermions serve as the foundational building blocks of matter, while the mind represents the smallest unit of intellect. The temporal nature of cognition denotes a perpendicular transformation of sensory reality. The temporal orientation endows the mind with its most peculiar qualities. Particles have a location in space, while thoughts and ideas exist in time. The 'Self' is a self-contained, holistic mental reality, forming a particle-like isolation. Nevertheless, its energy levels are far lower than the immense energies elementary fermions represent.

Considering the fermionic structure as an energy unit indicates that the particle structure might be valid for the universe as the originator of all organization and energy. Therefore, matter fermions (like electrons), the mind, and the cosmos could be considered three passible elementary particle categories. These entities' identical energy structures and analogous operational principles create a fractal arrangement encompassing various energy levels and sizes. The tiniest constituents are material fermions, characterized by immensely high frequencies and substantial energy levels. They constitute the radiant cores of stars and the delicate forms of living organisms. The largest cosmologic structure, the universe, possesses the lowest energy level, yet it houses infinite complexity within its finite confines.

The notion that the mind is integral to a harmonious, interconnected universe aligns with ancient perspectives. Aristotle, for instance, believed that the essence of an object existed within the thing itself. In Hinduism, the concept of Atman portrays the individual Self as one with Brahman, the Ultimate Reality and divine spirit of the universe. In monotheistic religions, God is synonymous with the cosmos, and humanity is fashioned in His image. We will explore the implications of the temporal fermion in individual and social behavior.

Classification of Emotions

Performance is enhanced through feedback from the environment. The initial language that the mind becomes acquainted with is not the native tongue but the physics vocabulary; even within the womb, muscular movements offer a sense of weight and proportion. The answer to the question, "What is it like?" is the privilege of emotions. Emotions can convey imbalance, thereby assisting in maintaining a state of restful equilibrium.

How can individuals make complex assessments in real-time? Our reality is structured by the four elementary interactions:[16] gravity,

---

[16] Fundamental interactions of nature govern behavior and decay of elementary fermions.

electromagnetism, and the nuclear weak and strong interactions. These interactions are facilitated by bosons, which transfer energy and information among matter particles, while emotions transmit inspiration. Emotions, the foundational driving forces of motivation, are the temporal analogs of physical interactions. Subsequent sections delve into the societal parallels of the most familiar interactions, gravity, and electromagnetism. We shall commence by examining the concept of gravity.

Gravity

"Spacetime tells matter how to move; matter tells spacetime how to curve." — John Archibald Wheeler

Despite being the weakest force in physics, gravity exerts pressure on every single molecule within our bodies. As articulated by Wheeler's renowned words, particles choreograph the dance of the gravitational field,[17] steering the motions of galaxies, stars, and planets. Objects of considerable mass induce more pronounced curvature in the fabric of space; even on Earth, gravity is stronger near mountains than in the less dense oceans. Consequently, substantial gravitational disparities exist on the scale of galaxies and galaxy clusters.

While we often take it for granted, gravity enables us to gauge distances, coordinate our movements, and perform abstract thinking. The intuition of gravity holds fundamental importance across the spectrum of biology. What can our exposure to gravity teach us about our own psychology?

Impact on our Emotional State: Although intense gravity can be perilous, the subtle fluctuations in its force evoke profound pleasure. Roller coasters around the globe captivate thrill-seekers by manipulating the influence of gravity. The ascending motion of the ride compresses the body, concentrating the pressure sensation around the stomach. Upon reaching the pinnacle of the ride's path, the dominance of gravity gives way to the exhilaration of weightlessness, engendering a feeling of expansiveness. Gravity influences our emotional states; from infancy to old age, we delight in swings, cradles, rocking chairs, parachuting, and bungee jumping. As gravity permeates our psyche, it endows us with a bipolar regulation of emotions.

---

[17] Fields are expansive, undulating waves of energy, generating a specific force at every point of space.

Emotional (Temporal) Gravity

"Holding on is believing that there's only a past; letting go is knowing that there's a future." — Daphne Rose Kingman

The lessons learned extend beyond mere logic when a child takes a tumble. Expressions like "above all", "parallel", and "distant" serve as linguistic indications that spatial orientation holds a place within the theoretical realms; physical attributes such as weight, size, and stability are projected onto mental constructs. The equilibrium found in physical balance is abstracted into a mental state of equilibrium. The heaviness of challenges and suffering, the pressure of stress, and the light moments of happiness originate in the experience of gravity.

Wheeler's quote acknowledges the concurrent development of particles and their corresponding fields. These fields also exist within the social sphere, intimately governing our emotional state and shaping our understanding (as depicted in Figure 8). Individuals within a network influence the social structure of their surroundings, thereby impacting the entire interconnected web (Li et al., 2018). The social environment, in turn, influences individual thought patterns, behaviors, lifestyle choices, and personal routines.

Like matter is held together by gravity, people are bound by relationships. Attachments to a spouse, child, parent, or a cause define

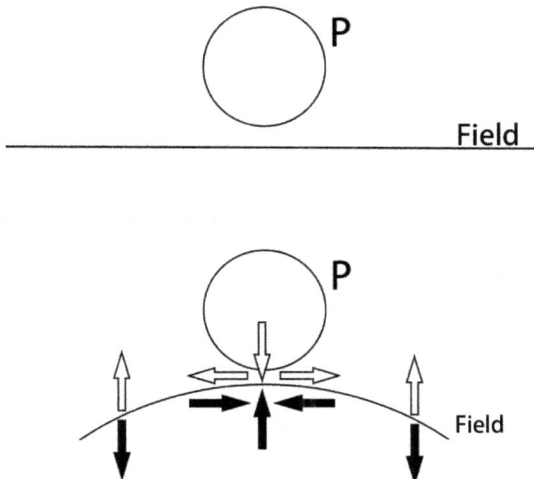

Figure 8. Interaction with the temporal field. 'P' indicates the mind 'Field' is the temporal field. Top: the mind is in harmony with its surroundings. Bottom: Sensory stimuli trigger activation, and response acts back on the environment. The evolution of the individual and its surroundings are intimately connected.

our loyalties, our secure bonds. The strength of these connections defines emotional gravity.[18] Temporal gravity oscillates between the yearning for security and nourishment and the allure of novelty (as illustrated in Figure 7). An emphasis on safety can lead to fear and conservatism. At the same time, an infatuation with the new can result in a sense of emptiness. Emotional gravity provides a stable foundation for our social lives, yet it also magnifies the pain of losses. It shapes our personal and social boundaries, molding our habits that impact every facet of our daily existence. Just as excessive or inadequate gravity is detrimental to the body, extreme psychological stress strangles the mind, while overindulgence weakens determination.

Because the thrill of levitation on roller coasters and swings is preceded by the pressure of gravity, the struggle possesses profound transformative potential. Challenges give rise to growth, turning individuals from victims into deserving heroes or heroines. Whether in folk tales, adventure narratives, horror stories, cliffhangers, or suspenseful accounts, hardship consistently paves the way for success. Just as periodic physical exercise is vital for the body, intermittent periods of stress benefit the mind.

Temporal Orientation

Fundamental to bodily function and perception are spatial orientation and balance. Within each ear, three semicircular canals are oriented perpendicularly to each other and function as a gyroscope,[19] sustaining posture even without visual cues (as illustrated in Figure 5). Stemming from our experience of physical space, mental orientation supports psychological stability and facilitates social and conceptual adaptation (Figure 3).

The temporal organization of the brain is rooted in the limited sustenance from essential factors like air, water, rest, and nourishment (Tsao et al., 2018), as well as the sequential associations crucial for learning, speech, thinking, and muscle coordination (Ahmed et al., 2020)—refer to Table I. Biological reliance on air, water, and food renders time the ultimate boundary of life. Our biology harmonizes through millisecond coordination, accomplished through a fluid, rhythmic interplay of glances, gestures, and smiles between caregivers and children, educators and pupils, supervisors and subordinates, and even strangers. Additionally, countless environmental cues swiftly prompt our spontaneous actions throughout the day. Similar to how material systems adhere to the

---

[18] Emotional (temporal) gravity is a social pressure, providing security and stress. At its extreme, it can cause a paralyzing tightness.

[19] A gyroscope has a spinning disc or wheel mounted on a base, which maintains its orientation regardless of any movement of the base.

Table I. The orthogonality of material fermions and the temporal mind. The mathematical formalism of quantum mechanics describes both systems. Still, their opposite high entropy manifestations mean a disorder in material systems but order and intellect for the mind.

	Matter fermion	Temporal fermion (the mind)
Unity	The smallest unit of matter	The smallest unit of intellect
Quantum mechanics	Applies in space	Applies in time
Entropy	The arrow of time	Future orientation
High entropy	Disorder	Order, creativity, intellect
Physical existence	Space	Time
Corresponding field	Gravity	Social manifold
Lorentz transformation	Time dilation	Dilation of time perception
	Gravity	Stress (emotional gravity, negative emotions)
	Acceleration	Mental expansion (awe and other positive focus)
Pauli exclusion principle	Forms material structure	Causes social hierarchy
Thermodynamic outcome	Exothermic cycle	Endothermic cycle
Particle-like stability	Particles are unchanging	A constant sense of self throughout life

principle of least action[20] (stationary action) when moving through space, cognition follows a trajectory of minimal temporal distance between the past and the future.

Through repeated use, specific neural networks become more dominant, essentially becoming 'hardwired' and significantly easier to activate. When confronted with decisions, the well-worn pathways within our brain offer the path of least resistance, forming a stationary route. Consequently, our current actions contribute to shaping our future social milieu.

Beliefs accumulate through experiences, fostering enduring mental stability. Memory endows intellect with predictive capabilities, anticipating environmental shifts and influencing the future (Northoff et al., 2019). In this way, the mind functions as a temporal compass, imbuing our life with direction by spontaneously navigating between birth and death. Subsequently, we delve into the exploration of how the brain gives rise to the psyche.

---

[20] According to the principle of least action events follow the minimal energy expenditure path.

The Role of Temporal Orientation in Perception

The arrow of time demonstrates a profound link to the rate of entropy generation (Lucia and Grisolia, 2020) and the loss of potential for work. Within non-equilibrium systems, the second law of thermodynamics elucidates the irreparable exchanges between states (Seif et al., 2021). The above irreversibility is analogous to the neural flow of activity within the neural system initiated by sensory input.

The challenge of neuroscience is how subjective mental states emerge from neural dynamics with a complex interconnection between the dimensions of space and time. Place cells within the hippocampus convert spatial relationships into a temporal projection (Shimazaki, 2020), thereby rendering temporal orientation a foundational aspect of the mind (as depicted in Figure 9). The notion of Self is inherently intertwined with temporal integration across various time spans, likely encompassing an individual's retrospective lifetime.

Spatial data compression engenders an orthogonal[21] sensory hologram (Saaty and Vargas, 2017) on the two-dimensional cortical surface (Déli, 2020a). Holography generates three-dimensional images, viewable from diverse perspectives. This is similar to the constant influxes of new experiences, which continuously reshape the significance of memories and one's sense of Self.

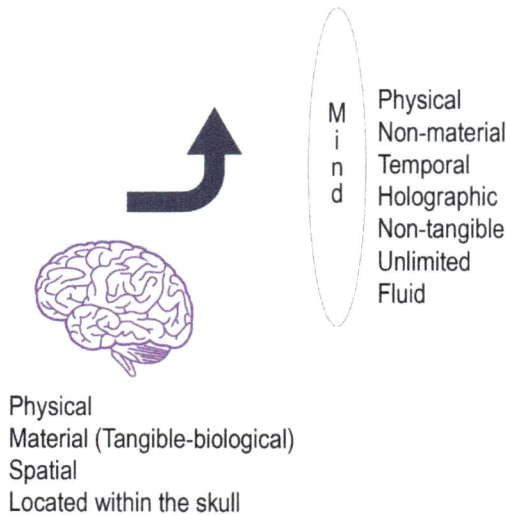

Mind

Physical
Non-material
Temporal
Holographic
Non-tangible
Unlimited
Fluid

Physical
Material (Tangible-biological)
Spatial
Located within the skull

Figure 9. The brain-mind axis. Sensory perception transforms spatial information into a temporal signal. The mind, a predictive organ, generates meaning and memory.

---

[21] Orthogonality is the quality of being perpendicular.

Subsequently, we explore how the perception of time shapes motivational dynamics.

Dilation of Time Perception

Our biological requirements and even social interactions oscillate between scarcity and surplus. We find joy in the arrival of relatives, yet we might be even more content when they depart. This shift between contentment and restlessness imparts a malleable quality to our subjective sense of time.

How does the psychology of emotions produce performance? The capacity of emotions to stimulate motivation and behavior underscores their energetic essence. Engaging in conversations with a romantic partner, a tax advisor, or an armed intruder demonstrates vastly distinct psychological states and varying rates of information flow. The pace of information transmission constructs a subjective perception of time, ranging between a scarcity of time and an abundance of it. Positive emotions and awe convey the sense of information voids, emerging from expansive feelings forming time excess. In contrast, stress parallels an overflow of information, leading to the sense that time is running out.

Our motivation is closely intertwined with our perception of available time.[22] A sense of awe stretches our perception of time, inspiring patience (Rudd et al., 2012). A feeling of permanence nurtures confidence (Makarevskaya, 2018), contentment (Carmona-Halty et al., 2019), generosity, and trust (Connelly et al., 2017). Conversely, stress acts as psychological pressure that constricts the flow of time (Remmers and Zander, 2018). The sensation of permanence makes every minute unbearable, resulting in impatience, impulsivity (Hosseini Houripasand et al., 2023), and the activation of the fight-or-flight response (as depicted in Figure 10; see Table II). It's important to note that it's not the negative emotions themselves but the transitions from neutral to negative states that induce the stagnation of perceived time (Wang and Lapate, 2023). Consequently, perception of time signal energy shifts during neural processing.

Warped spatial patterns across the brain (Knapen, 2021) can influence temporal curvatures, dimensionalities (Tozzi et al., 2017; consult Table II), and neural efficiency (Velasco et al., 2019). The boundless energy of positive emotions and the immobilizing constriction of negative feelings represent the polar opposites of our psychology. Thus, the dilation of time perception during positive and negative states reflects opposing energetic conditions and attitudes. As awe induces relaxation and extreme stress causes paralysis, neither catalyzes effective action.

---

[22] Time perception is the subjective experience of the length of events. Both negative and positive emotions lengthen time perception (Lupien et al., 2007).

Figh-or-flight response

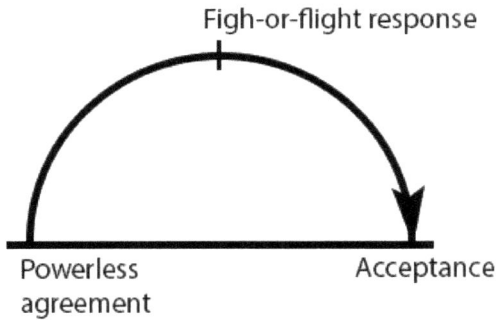

Powerless   Acceptance
agreement

Figure 10. **The polarity effects of temporal gravity.** Negative emotions can deplete energy and trigger the fight or flight response; however, at their extremes, emotions can result in no action.

Table II. **The physiological consequences of different brain states.** The thermodynamic and psychological implications of basic emotions.

	Reversed Carnot cycle High entropy resting state	Carnot cycle Low entropy resting state
Mental state	Positive emotions	Negative emotions
Frequencies	Oscillations expand (information poor)	Detail-oriented oscillations contract, accumulate information
Temporal dimensionality	Negative temporal curvature (higher dimensionality)	Positive temporal curvature (dimensionality loss)
Subjective sense of time expands	The wealth of time inspires confidence	Time pressure causes impatience
Future degrees of freedom	Expanding degrees of freedom	Degrees of freedom contracts
Thermodynamic consequences	An endothermic cycle absorbs energy and entropy from the environment.	An exothermic cycle dumps energy and entropy onto the environment.
Consequences for the organism	High mental energy (intellect)	Degradation of mental energy (mental and immune problems)

How does time perception relate to gravity? The memory of the field resides within its curvature. Within general relativity's framework, gravity and acceleration lead to time dilation. As gravity warps space, time passes more slowly at the base of a mountain. Similarly, the curving temporal field decelerates time perception (as illustrated in Figure 8). Emotional gravity modulates data flow between time pressure (i.e., stress) and expansion (i.e., elation), deviating from the Euclidean field formed by the brain's resting state. Although both conditions expand time perception (Figure 5), their divergent energy processes result in distinct physiological and psychological encounters and motivations. The deceleration of time perception in contrasting conditions underscores the brain's adherence to the principles of physics.

Emotions and Temperature Regulation: The brain's emotion and temperature regulatory pathways are closely linked through the striatum, limbic structure, and cortical insula (Inagaki et al., 2019). Therefore, the body's temperature control mechanisms can be utilized for psychological regulation through sweating or shivering (Kuno, 1956). For example, anxiety disorders often intertwine with disturbances in thermoregulation (Raison et al., 2015). Therefore, in addition to time perception, secondary emotional symptoms, such as shivering, sweating, changes in facial skin coloration, and others, enhance motivation.

The Social Mind

"For a solitary animal, egoism is a virtue that tends to preserve and improve the species: in any kind of community it becomes a destructive vice." — Erwin Schrodinger

From the perspective of Abraham Maslow, the American psychologist, the quality of our social environment plays a paramount role in our psychological well-being. The framework for our actions is established by physical laws, legal regulations, traffic rules, and social norms. Undoubtedly, our operational reality emerges from a foundation rooted in the social realm, forming a stable system of beliefs—akin to a floor. Conversely, limitations act as constraints, restricting possibilities like a ceiling. Consequently, our scope for movement is confined within the conceptual space defined by this floor and ceiling.

The curvature of the gravity field compresses physical space, while the curvature of temporal gravity gives rise to stress. Similar to how photons disperse energy throughout the cosmos, emotions disseminate motivation across society. The feedback to the mind's spontaneous exploration concerning our physical existence, behaviors, and social interactions unveils our standing within the social landscape. For instance, playful teasing among young boys establishes dominance dynamics. Group members thoroughly understand their social positions and vigorously safeguard them. Our perceived rank within the social hierarchy shapes our aspirations, behaviors, decision-making processes, and countless subconscious reactions in our daily lives. These influences give rise to social hierarchies, classifying socioeconomic strata according to factors like wealth, income, race, education, and authority, as discussed in the "Conservation of Time". As our social environment molds our ambitions, we simultaneously shape and originate our world.

The Structure of Society: Socioeconomic Status

"Our inequality materializes our upper class, vulgarizes our middle class, brutalizes our lower class." — Matthew Arnold

Fermions' distinctive half-integer spin engenders atomic arrangements and an intriguing resilience to compression. Consequently, matter, including massive stars, resists compression, forming the sturdy foundation of our reality. The fermionic framework's resistance to stress leads to the territorial behaviors of animals. Similarly, human beings safeguard their domains of influence, leading to the establishment of hierarchical social structures (as outlined in Table I). Social representation can be likened to a topographical landscape within the conceptual social field, where influential individuals and organizations occupy the peaks while ordinary people crowd the plains.

It enlarges social conflicts, increasing the need for policing; it devastates human potential, negatively affecting technological progress.

Human civilization has advanced toward notions of social mobility and democracy, yet hierarchical relationships endure and perpetuate inequality. Higher social status often aligns with a dynamic environment and opportunities, fostering optimism and individuality. A positive mindset nurtures confidence, adaptability, and intellectual growth. Trust and faith serve as sources of spiritual strength and prerequisites for intellect. However, social status is an artificial category that discourages genuine relationships. Therefore, inequality within and among nations proves detrimental to progress and long-term stability. It exacerbates social conflicts, necessitating heightened policing, and curtails human potential, hampering technological advancements.

Open social frameworks should ideally present avenues for achievement. However, despite the potential for equal opportunities, why is inequality rising in many parts of the world? What distinct challenges confront the impoverished? In a manner reminiscent of gravity forming dense structures with maximum pressure at the center, social gravity generates stress directed at the people with the least social financial resources. Hierarchic social structures redistribute time among people, with the access to time being inversely proportional to social status, imposing the most significant stress on the marginalized (Figure 11). The restricted access to healthcare, justice, and education diminishes life expectancy; brutality, injustice, and indecency contaminate the human spirit, leading to mental degradation, abuse, and conflict. The burden of intergenerational anxiety drives individuals into a sense of hopelessness, which can ignite acts of terrorism and crime.

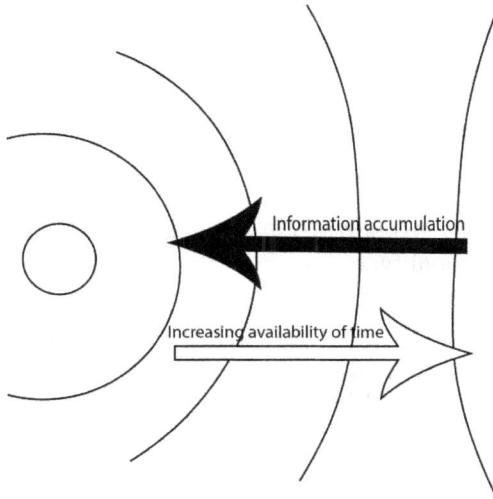

Figure 11. Social status as a function of temporal curvature. Competition results in an uneven distribution of time, reducing it for those on the left and increasing it for those on the right. The accumulation of information is stressful due to a positive temporal curvature (as indicated by the black arrow), whereas greater access to time leads to satisfaction and confidence (white arrow). Lack of time on the far left provokes a struggle for survival.

Temporal Mass

"You only lose what you cling to." — Buddha

In physics, mass quantifies inertia, denoting the resistance to an applied force. Similarly, temporal mass (or emotional mass[23]) gauges the intensity of attachments. Strong attachments impede social mobility as heavy objects settle and resist change.

Low Temporal or Emotional Mass: Low temporal or emotional mass signifies a limited capacity for decisive action. Factors such as wealth, poverty, or affiliation with religious institutions can isolate individuals from active social engagement, resulting in ignorance, represented by low emotional mass. On the other hand, moderate temporal mass establishes a stable intellectual foundation, enabling the development of a sense of purpose.

Large Temporal Mass: A high temporal mass is associated with intense attachments. Though not a financial classification, poverty can intensify

---

[23] Temporal mass is the strength of attachments to things and people. Emotional mass provides security but enhances attachments, which slow change and block progress.

temporal mass. This substantial temporal mass fosters insecurity, which in turn fuels fear. Insecure individuals often employ complaints, excuses, and conspiracy theories to mask their insecurities and fears. This tendency to complain amplifies problems instead of solving them. Those who take action do not have the energy to indulge in complaints. Fear also triggers a compulsion for excessive control, hindering progress. A rigid adherence to traditions leads to beliefs in severe punishments. The condition can deteriorate into a glorification of suffering and self-flagellation, indicating a pathological personality. Such rigidity obstructs change and obstructs upward social mobility.

In the words of George Orwell, "Poverty…annihilates the future." Poverty imposes an unrelenting cognitive burden, diminishing intellectual capacity (Makharia et al., 2016; Mani et al., 2013). The pressing concerns with immediate needs muddle cognitive abilities. It erodes security and personal boundaries, creating an emotional barrenness that erases traces of love and engenders callous disregard for societal norms. Distrust perpetuates itself by extinguishing generosity, kindness, and civility. Expecting honesty from the destitute is futile; their moral compass is broken.

Psychological Lensing: Light follows the contours of space. Gravitational lensing, which results from the bending of light by massive objects, can distort the apparent location of an object or create mirages. Remarkably, this bending of light parallels the emotional distortion of memories and experiences. Psychological lensing is the emotional equivalent of gravitational lensing. Much like gravitational lensing challenges our perception of the cosmos, context-dependent bias obscures the truth.

Identification with a cause or a person (family member or even celebrities) can become central to people's personal or group identity. Not getting the outcome people want on something central to their identity causes them to let go of objectivity and discredit the process (Mullen and Skitka, 2006). This situation-dependent bending of the facts produces psychological lensing's alternative 'realities'. Sometimes, this bias is confined to a specific domain, but like cancer, it can infiltrate various aspects of life.

The Laws of Emotion Regulation

"The laws of nature are written in the workings of our brains."
— Thomas L. Saaty

The profound parallels between the mind and matter carry extensive implications for society. The bedrock principles of classical physics hinge on the stability of the gravitational field. At the same time, the societal sphere depends on social equilibrium. Newton's and thermodynamics principles underpin our physical reality and leave an indelible imprint on our psychological makeup. Universal gravitation governs the trajectories of celestial bodies, just as our aspirations take root in our societal strata. Consequently, classical physics' tenets find relevance within society's fabric.

Newton's Laws

Newton's trilogy of motion laws is the cornerstone of classical mechanics, delineating how forces influence motion. Their temporal projections reveal the deterministic potency of emotions.

The First Law

The law of inertia dictates that objects persist in a state of rest or uniform motion. This law finds temporal counterparts in emotions and habits. Emotional inertia sustains prejudices, grievances, and animosities across generations. Adjusting to life changes or contending with losses is impossible without emotional investments. Consequently, embracing an unsatisfactory situation often proves more attractive than unsettling the comfort zone.

The challenge of creating positive habits and breaking negative ones has motivated various institutions, including schools and the legal system. Prisons aimed at reforming inmates inadvertently foster impulsiveness— the root of criminal behavior (Umbach et al., 2018). Newton's first law suggests that addressing underlying causes instead of symptoms is pivotal, underscoring the importance of preserving individual dignity.

Like a fledgling chick emerging stronger after endeavoring to hatch from its shell, efforts invested in forming positive habits, studying, accepting, and meditation instill confidence for ongoing mental advancement. The "Free Will" section delves into habit-related discourse further.

Newton's Second Law

A constant force yields acceleration. The second law emphasizes that social influences are additive in the societal realm. Genuine compliments can propel significant personal growth, whereas maltreatment compromises psychological well-being. For instance, skilled educators, coaches, and counselors can positively influence generations of young people under their tutelage. Conversely, the aforementioned prison scenario illustrates how the degrading and humiliating atmosphere reinforces the very qualities it should rectify.

Newton's Third Law

Newton's third law establishes the symmetry of forces: two bodies exert forces on each other with equal magnitude but opposite directions. A body maintains its state unless compelled by external force.

The Man of la Mancha echoes this symmetry of forces:[24] "Whether the stone hits the pitcher or the pitcher hits the stone, it's going to be bad for the pitcher." Carl Jung, the Swiss psychiatrist and psychoanalyst, posited that a compensatory unconscious mirrors our lives. Each conscious action engenders a corresponding and opposing unconscious reaction, striving to balance the human experience.

The social dimension of this law delineates the source of social discord. A reactionary force equals the force exerted—people push back as hard as we push them. For example, innovative concepts spark unyielding opposition. The reactionary forces can take various forms—everything from active resistance to passive-aggressive behavior to sabotage behind the scenes. It's also important to realize that this opposing force is not always negative—it can provide opportunities to engage in quid pro quo to achieve mutual goals.

Third Law situations move toward a state of equilibrium, in which equal forces are exerted against each other. Opposing social forces sustain a delicate balance: conservatives and progressives, left and right, protectionism and globalization epitomize political reflections of one another. Their apparent adversarial conflict is but a façade. In truth, the existence of one hinges upon the presence of the other. Polarization always has two sides.

The Laws of Thermodynamics

"Regarding the Laws of Thermodynamics: "(1) You can't win, (2) you can't break even, and (3) you can't get out of the game."
— Dennis Overbye

Thermodynamics delves into the conversion and transfer of energy. Einstein succinctly encapsulated its significance: "The laws of thermodynamics are the only physical theory of universal content which I am convinced ... will never be overthrown." Applying the four laws of thermodynamics to societal contexts yields profound insights into the nature of social existence and the foundations of society.

---

[24] Man of la Mancha, musical, 1965.

The Zeroth Law of Thermodynamics

The Zeroth Law asserts that if two thermodynamic systems attain thermal equilibrium with a third, they are in thermodynamic equilibrium. This concept defines temperature, which aligns with thermal energy in physics. Just as particles' kinetic energy defines temperature, the tendency for competition outlines social temperature (Déli, 2020a).

(1) Low social temperature epitomizes predictability, trust, and civility. In the late 1800s, Peter Kropotkin observed that abundance fosters cooperation and generosity among wild horse herds, wolf packs, and human communities (Connelly et al., 2017; Kropotkin, 1902). Thus, social collaboration and generosity hinge upon societal safety and equity.

(2) As resources diminish to a critical point, generosity yields to competition. Stressed, strained ecosystems give rise to chaos and disorder, producing high social temperatures. Social biases and oppression arise as a natural outcome of perceived scarcity. Resource scarcity imposes a cognitive burden, hampering IQ.

Leo Tolstoy astutely cautioned, "There are no conditions to which a man may not become accustomed, particularly if he sees that they are accepted by those around him." The Zeroth law finds application in social climates through the bandwagon effect, where newcomers swiftly adopt local social norms. Akin to this, healthy rats, pigs, and humans can acquire depression from afflicted peers in prolonged cohabitation (Zeldetz, 2018).

In the nineteenth century, Thomas Malthus noticed that food production did not keep pace with population growth, causing perpetual poverty, war, famine, and disease. Paul Ehrlich's book, *The Population Bomb* (1968) brought the question into the public consciousness. None of these threats came true. With advances in agricultural technology, seeds, soil management, irrigation, and mechanization, the food supply stayed ahead of the population curve. Nevertheless, humanity's future hinges on finding sustainable growth strategies that respect Earth's resources and the natural environment. The Social Zeroth law reminds us that only a shared future is viable for humanity.

The First Law of Thermodynamics

The law of energy conservation stipulates that energy cannot be created nor destroyed — only transformed from one form to another. Put differently, the total energy within an isolated system remains constant. Exothermic processes dissipate entropy and energy into the environment, making endothermic systems that absorb entropy and energy possible.

Conservation of Time: We have seen that the orthogonal transformation of spatial input is crucial in the temporal organization of cognition and intellect. Our psychology hinges on the preservation of time. Time conservation originates in the time needed for securing sustenance and evading predators. Tomorrow cannot compensate for today's lack of food, air, and water. Life's finite nature underpins this principle. In the realm of society, technological progress shapes our temporal allocation. In medieval times, nearly 90% of individuals toiled in agriculture for collective sustenance. Presently, this figure stands at 10% and is set to decline further, freeing a significant portion of society to engage in diverse pursuits.

As earlier elucidated, the least action in physics resurfaces in cognition as a stationary temporal trajectory. Time takes the place of space in social interaction, and energy conservation becomes time conservation. While technology has drastically diminished the time required for survival needs, uneven access to time engenders psychological divisions. Power affords relaxation, while inferiority generates stress. Inequality is fundamentally the inequality of access to time. The affluent can purchase time, as the less privileged are willing to trade their time for money. The dearth of time for the disadvantaged finances the excess time of the affluent, lending credence to the adage, "Time is money" (depicted in Figure 11). Consequently, the comfort at the apex relies on the hardship of the many, rendering social mobility an illusion. Time constraints make escaping dire poverty through diligent effort more unlikely than winning a lottery.

The Second Law of Thermodynamics

The Second Law elucidates the inclination of isolated systems to degrade into disorder, with systems naturally progressing toward thermodynamic equilibrium characterized by maximum entropy.

In physics, entropy is often characterized by the microstates of the system and the different possible arrangements of particles. Although not directly transferable to societal contexts, individual actions can yield comparable state formations. Continuous sensory-driven pairwise comparisons create an "inner eye" (Peng and Xie, 2016; Saaty and Vargas, 2017). Comparison with others generates peer pressure that levels opportunities within a given class and disseminates habitual behaviors, such as quitting smoking, opinion forming, experiencing happiness, and adopting optimism or pessimism (Kong, 2022).

The concept of sentiment contagion[25] gauges the likelihood of emotional transmission among users in online social media. Our intellectual and

---

[25] Sentiment contagion is the ability of sentiments and behaviors to spread like infectious disease in social groups.

social attitudes are universal, unconscious, and convergent, homogenizing intentions and opinions into a shared intuition (Goldenberg et al., 2021). Thus, sentiment contagion increases social entropy, analogous to fluid dynamics, as discussed in "Emotional Temperature and Pressure".

However, while competition fosters similarities among individuals with comparable economic backgrounds, it concurrently fosters social discord and inequality. Across 71% of the global populace, inequality has burgeoned, concentrating income and wealth at the apex. There is an expanding chasm in prosperity and opportunities, which warrants further exploration. In the context of physical systems, low entropy magnifies the system's capacity to generate work. Likewise, increasing social friction signifies societal instability, catalyzing uprisings and revolutions.

The Third Law of Thermodynamics

The Third Law posits that as a system reaches absolute zero temperature, its entropy approaches zero. Lower temperatures significantly decelerate the pace of interactions. Emotional temperature can apply the law to society. Civility is proportional to socioeconomic status but exhibits an inverse relationship with emotional temperature. As previously demonstrated, higher social status affords abundance and security (i.e., lower social temperature). Akin to physical systems, social entropy tends toward zero under low social temperatures. Serenity and mental balance render competition obsolete in secure environments (Kropotkin, 1902).

Just as physical systems possess vibrational energy even at absolute zero, some friction remains an intrinsic facet of any society. Lingering social energy renders complete non-reaction and equanimity implausible, even for the saintly individuals. Learning and progress are contingent on interactions and mistakes.

Emotions constitute an inherent component of the human experience and are inherently intertwined with society.

Emotional Electromagnetism

"The starting point of all achievement is desire." — Napoleon Hill

Just as gravity serves as the bedrock of space, emotional gravity underpins society. Much like the necessity for charge separation in atomic structures, the organization of society involves the coexistence of opposites, such as male and female, democrat and republican, conservative and progressive. While emotional gravity functions as a steady force, emotional electromagnetism can rapidly oscillate between attraction and repulsion (Drigotas, 1993). For example, a dog's wagging tail signals desire, while raised hackles convey aversion.

Polarization obstructs the pursuit of objective and impartial viewpoints. The act of excommunicating a member reinforces the unity of organizations or nations. Individuals with adversaries seek camaraderie, expecting their friends to share their anger or antipathy toward the so-called enemy (Krems et al., 2023). Hate demands substantial energy, possibly overshadowing life's rational equilibrium and depleting common sense, time, and financial resources. Recognizing this universal human susceptibility permits a better life balance.

Due to its greater strength compared to gravity, emotional electromagnetism can spur extraordinary accomplishments, extravagant purchases, and imprudent decisions. It also makes unsuspecting individuals susceptible to manipulation by demagogues and fraudsters. This phenomenon also surfaces in the disappointment in response to unexpected cancellations. Just as flipping the poles of a magnet alters its attraction, a shift in the social climate can instantly turn interest into antagonism and repulsion.

Emotional electromagnetism encompasses competitiveness, aggression, and audacity. These attributes, associated with masculinity, tend to predominate in youth. Throughout our lifetimes, the predominance of emotional gravity relative to electromagnetism tends to increase. In older individuals, emotional gravity safeguards against envy and resentment, although stronger attachments render them vulnerable to the experience of loss. Women also exhibit a stronger gravitational tendency, often displaying a natural inclination toward cooperation.

Positive Psychology and Expectations: Positive psychology acknowledges the significant impact of hope on personal growth. Expectations act like a magnetic field, substantially influencing those who embrace them. Children, for instance, frequently live up to the expectations set for them. However, it's important to note that Lenz's law operates on a deeper level than mere surface or wishful thinking; it involves genuine intentions grounded in beliefs.

Festinger's Cognitive Dissonance Theory

Constructing a false facade (analogous to a magnetic field) induces a corresponding emotional charge. Festinger's Cognitive Dissonance Theory illustrates how immoral behavior has a reciprocal effect on our core principles, inevitably leading to their alteration. Experiences conflicting with our beliefs cause a mental strain called cognitive dissonance. The discomfort may cause us to twist our perception of reality to reduce this discomfort or even change our beliefs.

Rigid moral rules (magnetic fields) can trigger a counterflow of evasive actions, deceit, and secrecy. This phenomenon is especially concerning in children. For instance, boasting and belittling can foster feelings of inadequacy, temptation can weaken resolve, delight can give rise to envy, displeasure can fuel paranoia, and pride can engender desire. However, projecting a fake social milieu can precipitate destruction.[26] Ultimately, bullies and tyrants become victims of their poison.[27]

Harmonic Motion

Emotions influence us around the clock, every day of the week. Feelings follow a pattern akin to harmonic motion; peaks generate attraction, while troughs elicit disdain, even within our closest relationships, such as those with our children or spouse (Figure 12). Similar to a spring's oscillation, the equilibrium point holds the most significant momentum, propelling attitudes to switch to their opposite. Because the equilibrium position hides feelings from awareness, we are ignorant of their alternating nature and power.

Emotions are under the control of the waveform's automatic evolution. Interaction can change the wave's amplitude, frequency, and intensity. Impulsivity increases the frequency of interaction, and aging reduces its amplitude. As discussed in later sections, the waveform can produce emotional interference or passive aggression. On a larger scale, intergenerational, organizational, and national sentiments also follow

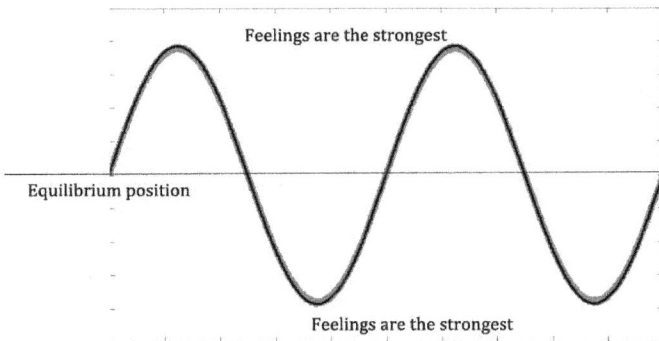

Figure 12. Emotional electromagnetism. Emotions can be represented by a harmonic function, with the most intense feelings located at the peaks and troughs of the curve. Emotions are hidden at equilibrium points. The highest point of the curve represents the strongest attraction, while the lowest point corresponds to intense dislike.

---

[26] Death of a Salesman, Artur Miller, 1949; Madame Bovary, Gustave Flaubert, 1856.
[27] Plato, The Republic.

harmonic motion. These forces turn repressions and revolutions into inevitable consequences of each other. The 1956 Hungarian uprising, the 1968 Prague Spring, the 2004 Orange and 2014 Maidan Revolutions in Ukraine, and the 2012s Arab Spring all grew out of brutal oppression and hopelessness.

Lenz's Law

Lenz's law applies the conservation of energy to electromagnetic interactions. When a change in magnetic flux through a loop occurs (such as the movement of a magnet or a difference in the current of a nearby circuit), an electromotive force (EMF) is induced in the loop. This EMF will generate a current in a closed loop. The law states that the direction of the induced EMF (and thus the direction of the induced current, if any) opposes the change in magnetic flux that produced it.

Decreasing Magnetic Flux: If the magnetic flux through the loop decreases, the induced current will circulate in a direction that tries to maintain or increase the existing flux. For example, if a magnet's north pole is pulled away from a coil, the coil will act as though it's producing its own north pole to try and keep the magnet from leaving.

Increasing Magnetic Flux: If the magnetic flux through the loop increases, the induced current will circulate in a direction that tries to reduce or counteract this increase in flux. If you push a north pole of a magnet into a coil, the coil will act as if it's creating its own south pole to oppose the magnet's entry.

Lenz's law says that the direction of the induced current will oppose the change in flux that created it. It's not a matter of increasing and decreasing; it's a matter of direction. A decreasing flux creates an EMF that creates a current, and that current will be in whatever direction it has to be to try to keep the flux from decreasing. An increase in flux will lead to a current that will try to keep the flux from increasing. In both cases, it's consistent with conservation of energy.

In our daily lives, Lenz's law keeps our mindset congruent with our social situation. When our thought processes and beliefs are disjoined, our mind tries to restore congruence. For example, losing social support urgently requires finding it elsewhere. Unfair treatment or disappointments often trigger a knee-jerk, revengeful, or destructive reaction. The so-called impostor syndrome is the self-sabotage of significant social advancement. The most secure way to achieve success is by gradually improving our social position.

Thought Suppression: Thought suppression involves a conscious effort to halt anxious or depressing thoughts. Lenz's law demonstrates that focused suppression is ineffective, allowing negative thoughts to resurface with heightened intensity, particularly during relaxation or fatigue (Roemer and Borkovec, 1994). Lenz's law also contributes to the challenge of altering habits or overcoming addictions.

In societies and organizations, the rhythm of the usual way of life provides stability. External changes trigger strong opposition, ranging from foot-dragging and inertia to outright rebellion, attempting to keep the emotional charge flowing.

Emotional Electromagnetism in Social Behavior

"The world is little, people are little, human life is little. There is only one big thing—desire." — Willa Cather

Electromagnetism is a foundational force in the physical realm, with magnetic fields spanning vast cosmic scales, even encompassing the universe's most significant clusters of galaxies. It follows that social bias (comparable to magnetism) is an important factor in shaping societal structures. Charge separation in the human context leads to biased and selective thinking, often manifesting as an "us versus them" mentality. Oxytocin, a hormone, can accentuate this division, promoting affectionate behavior toward familiar individuals while triggering hostility toward outsiders (Anpilov et al., 2020).

Attachments to family, social groups, nations, or races can blind individuals to the flaws of their idols and affiliations, whether in politics, sports, or religion. Emotional electromagnetism adeptly fosters group unity by vilifying scapegoats, whether individuals or groups. Leaders might enhance their popularity by stoking sentiments of racism or anti-immigration. Historical examples, such as communist and Nazi regimes, utilized ostracism to consolidate political control. Conversely, polarized populations often rally behind strong leaders.

However, unity built upon bias carries risks. Just as similar charges repel each other, in-group cohesion can be short-lived and result in internal conflicts. Divisions along political, religious, ethnic, or geographic lines often lead to hostility and fragmentation. Continuous feuds plague prejudiced and segmented communities, like the Montagues and Capulets, Christians and Muslims, or Sunni and Shia factions. Polarization and partisanship erode trust and hinder progress. Resilient and dynamic economies avoid significant polarization (Klimek et al., 2019).

With this, the classical exploration of the fermionic mind concludes. We showed that the brain uses any means necessary to reproduce analogous regulation to the material world. Consciousness is an abstract reflection of particle organization.

The upcoming section will delve into the quantum attributes of mental particles.

Chapter 4

Quantum Cognition

Quantum mechanics is a foundational theory in physics that describes the behavior of matter and energy on the smallest scales, typically at the level of atoms and subatomic particles. Unlike classical mechanics, which predicts deterministic outcomes, quantum mechanics is inherently probabilistic. For example, particles can exist in a combination of multiple states simultaneously. Only when we measure them can they 'collapse' into one of the possible outcomes. The wave function collapse means particles can behave like waves (with interference patterns) and like particles (with discrete energy levels). When squared, the wave function provides the probability of finding a system in a particular state.

Another quality is entanglement occurring between two particles. The state of one instantly affects the state of the other, no matter the distance between them. Einstein famously described this phenomenon as "spooky action at a distance". Furthermore, particles can 'tunnel' through barriers that classical mechanics predicts should be impenetrable. Quantum mechanics has been experimentally verified to an incredible degree of precision. Yet, debates about its interpretation continue to confound our conception of reality.

Thermodynamic Considerations of Brain Activities

Fresh and unforeseen information possesses the ability to startle individuals and societies alike. Emotional gravity provides our convictions for our secure mental world firmly rooted in resting stability. Unexpected disruptions give way to cognitive disarray, which compels individuals to explore ways to stabilize their beliefs. However, this process inherently reshapes the very framework of those beliefs. The more astonishing the news, the greater the mental exertion and time required to interlace those objective realities into the existing certainties. The intellectual labor and painful assessment needed to assimilate jarring information into the domain of convictions is a testament to the energy needs of cognitive transformations.

The Thermodynamic Cycle in Physics: In physics, a thermodynamic cycle describes various heat and work transfers due to temperature, pressure, and volume changes, where the system eventually recovers its initial state. A prime example is the Carnot engine,[28] operating between two reservoirs that define the maximum possible efficiency of a heat engine when converting heat into work. The Carnot cycle is a closed system[29] drawing heat from a hot reservoir and releasing it into a cold one, executing work without generating entropy.

An essential facet of the Carnot cycle is its reversibility, enabling the execution of both exothermic and endothermic processes. Exothermic operations in physics yield heat, whereas the endothermic counterpart necessitates an infusion of energy. Notably, the cycle's heat absorption or work production can be significant as the function of the temperature change.

The Brain as a Closed Thermodynamic Cycle: Biological organisms are open systems that consume energy-storing molecules and release metabolites. However, because the brain only interacts with the environment through the sensory and motor systems, it represents a closed system reliant on the resting state with constant parameters (Crossley et al., 2023). Therefore, the resting state functions as a thermal reservoir. At the same time, sensory perception constructs a self-contained thermodynamic cycle (as reviewed by Watts et al., 2020). In this manner, perception circles between the environment and the high entropy resting state.

The Brain's Energy Turnover: According to von Helmholtz (1882), the maximum work a thermodynamic process can perform on a constant volume equals the negative change in the free energy. Neurons within the brain expend energy to uphold their resting potential, activate action potentials, and transmit signals across synapses. The mitochondria manufacture adenosine triphosphate (ATP) from blood glucose, serving neurons' high-energy demand. This energy turnover is so efficient that heightened local activation avoids exhausting ATP reserves and augments them. As a result, increased blood flow and oxygen supply around the active cells can be used as a marker of local brain activity.

Complexity Generation: The brain exhibits an exceptionally high energy consumption rate, accounting for about 20% of the body's total energy usage. This energetic expenditure primarily sustains the brain's internal

---

[28] A theoretical thermodynamic cycle proposed by the French engineer Sadi Carnot.
[29] A thermodynamically closed system conserves mass but energy can enter or exit freely.

workings, with a smaller percentage allocated to activities evoked by external tasks (Huang, 2019). Consequently, the brain's intrinsic activity, occurring even without specific tasks, surpasses those prompted by stimuli.

In ordinary circumstances, the brain's heat exchange with the external environment remains minimal; energy and information transfer exclusively transpire through the sensory organs and the muscular system. Analogous to how a mechanical engine requires lubrication to operate smoothly, the brain's energy supply ensures the seamless transmission of information through its intricate neural network. However, in contrast to the work produced by a mechanical engine, perception by the brain gives rise to complexity.

Information Transmission: One of the brain's primary functions is information propagation. All memory storage devices, from the brain to the computer, store information by altering physical properties through energy-consuming processes. Significant energy consumption ensures the operational readiness of individual neurons (Fry, 2017). Notably, the speed of information transmission is contingent on the frequency of brain oscillations—higher frequencies transmit more information per unit of time than lower frequencies. Moreover, neurons cannot generate greater activation; instead, the firing rate of neurons reflects the intensity of stimulus and the speed and strength of muscle contractions. Therefore, stimulus imposes energy cost for the sensory input through the oscillation frequency.

The Reversibility of the Perception Cycle: The brain's treatment of sensory information bears semblance to how fuel operates in an engine. Much like a spark ignites compressed gases in an engine, triggering combustion and the subsequent piston-driven motion that propels the vehicle, the subjective significance of sensory information acts as the spark igniting the cognitive cycle. As a result, the meaning attributed to information determines the cycle's energy and direction, whether it's an endothermic or exothermic process. The intricate geometry of the cerebral cortex gives rise to a dynamic intrinsic activation pattern on its surface, akin to fluid waves (Pang et al., 2023). This pattern holds more influence over arousal than the nature of the stimulus itself (Dempsey et al., 2022), effectively rendering arousal a probabilistic and reversible thermodynamic cycle (Figure 2).

The Energy Consequences of Perception: Perception and memory formation drive oscillations beyond the constraints of the connection map (Inzlicht et al., 2018), thus fueling energetically demanding changes in synapses (Crossley et al., 2023). Cognitively taxing exothermic cycles that dissipate energy impairs cognitive functioning (Padamsey et al., 2022). Conversely, the endothermic process of the reversed cycle augments mental energy levels.[30]

The Quantum Characteristics of Cognition: The reversibility of the brain's cognitive cycle turns perception probabilistic. When the process reverses its direction, it creates an opposite thermodynamic outcome, lending a spin-like quality to perception. Furthermore, the cycle's dependence on stimulus introduces contextuality, and the switch between the irreversible resting state and the reversible cognitive cycle is analogous to the wave-particle duality. Therefore, the highly complex neural system adopts quantum operation.

Therefore, intelligent processing is analogous to information erasing, which decreases the temperature while increasing the overall neural organization (O'Neill and Schoth, 2022).

The Path to Quantum Cognition

"Life is strong and fragile. It's a paradox... It's both things, like quantum physics: It's a particle and a wave at the same time. It all exists all together." — Joan Jett

The dawn of the twentieth century marked a flurry of intellectual endeavors that would lay the foundation of quantum mechanics. However, the groundbreaking experiments, exemplified by the counterintuitive nature of wave-particle duality, perplexed and disconcerted physicists. Interestingly, an unexpected source of inspiration emerged, Carl Jung's analytic psychology, which shaped Niels Bohr's explorations into particle-wave duality. Furthermore, dialogues and exchanges with Carl Jung granted Pauli an almost mystic insight into particle physics.

Where did such unlikely parallels come from? Jungian psychology views the images and symbols that emerge within dreams and perceptions as offering insights into the structural underpinnings of the mind, similar to scattering experiments unveiling the atom's inner workings.[31] Beyond scattering experiments, various frameworks of physics can elucidate enigmatic cognitive phenomena. For example, quantum mechanics,

---

[30] The ability to transform information into energy, i.e., intellect, seems to be the brain's essential quality. The energy gain of the neuronal system is intellectual evolution, which is closely intertwined with the environment. Although the environment controls the brain's sensory input, the mind can control the environment.

[31] In physics, scattering experiments measure the deviation in the trajectory of particles following a collision with another particle. Scattering need not necessarily involve direct physical contact; instead, it can transpire through the interaction of particles repelling each other, such as in the case of two positively (or negatively) charged ions.

which captures particles' wave-like behavior, also provides insight into the context-dependence of decision-making. Like indivisible quantum entities, beliefs cannot be decomposed into smaller units. The main topics of human thoughts, such as God or politics, come down to affirming or opposing the idea. Beliefs are also updated through a discrete mental energy change, indicating the quantized nature of consciousness.

Quantum mechanics[32] has given us a wealth of technological advancements spanning atomic clocks, holography, Lasik Eye Surgery, smartphones, and quantum computers. Yet, beneath the impressive accomplishments over the past century, hide our complete ignorance of an intuitive understanding. Analogous to a present-day stomachache originating from tomorrow's meal, the peculiarity of quantum phenomena renders outcomes contingent upon far-off observers. As quantum mechanics challenges our firm grip on reality, might the field of cognitive science offer remedies for this conundrum?

Conscious perception, encompassing a state of wholeness, becomes subject to non-deterministic, quantum-like fluctuations between dual possibilities. Where does the quantum character originate? Is it our neurons or cognition itself that produces these intricate quantum psychologies?

Much like electrons or photons, the mind defies classical assessment. The mental state remains elusive until the moment of decision-making, at which juncture potentialities crystallize into concrete attributes. Introducing sensory information, even as simple as opening the eyes, disrupts and changes the brain. However, measurement (stimulation, particularly pronounced incentives) cannot establish motivation, for it irrevocably alters emotional states, rendering the original condition uncertain.

Just as an electron cloud within an atom delineates a specific energy level but not an exact position, the brain's response to identical stimuli varies slightly due to the influence of synaptic memories. Cognitive outcomes are intrinsically linked to the context of memories and situations. Quantum probability theory offers an explanatory framework for noncommutativity and order effects in decision-making (Pothos and Busemeyer, 2021). Additionally, Bose-Einstein statistics[33] can account for the bandwagon effect[34] in psychology (Khrenikov, 2020). In the digital age, social media platforms epitomize internet-based echo chambers, where coherent, high-amplitude informational waves can swiftly amass immense motivational energy.

---

[33] In bosonic system we cannot differentiate between the particles.
[34] The bandwagon effect is the *tendency for people to adopt certain behaviors, styles, or attitudes simply* because others are doing so.

Everyday experiences feature objects with discernible speeds at specific points in space; however, particle measurement harbors inherent limitations. Simultaneous observation of complementary physical properties, like position and momentum, remains unattainable. Yet, as probabilistic quantum attributes coalesce into predictable physics, individualistic actions form stereotypical social patterns. The mathematical principles underpinning quantum physics leverage probabilistic datasets in psychology to forecast preferences in music, purchasing decisions, and political choices.

Unity in Quantum Mechanics

In quantum systems, the potentialities encapsulated within a wave function manifest as amplitudes that eventually resolve into well-defined observational outcomes (Selesnick and Piccinini, 2018). Similarly, conscious perception exhibits an unbroken continuity: when confronted with ambiguity, a non-deterministic, quantum-like fluctuation transpires between two options, images, ideas, or concepts, as exemplified by the perceptual alternation in the Necker cube's dual depth representation (Einhäuser et al., 2008) see Figure 13. This phenomenon extends to binocular rivalry, where perceptions oscillate between the two sensory fields of the eyes (Tong et al., 2006) or the nostrils (Zhou and Chen, 2009).

Surgically separating the two hemispheres causes the famous split-brain experience. This division gives rise to disconnected mental

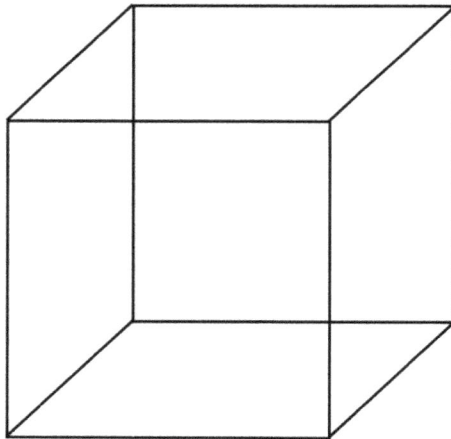

Figure 13. Necker cube. The Necker cube is a wireframe cube with no depth cues. We perceive the image as three-dimensional with either the lower-left or the upper-right square as its front side.

content, intentions, and distinct personalities within each hemisphere (Pinto et al., 2017). Intriguingly, a further partitioning of the brain leads to the annihilation of consciousness. Since particles represent matter's fundamental constituents, it's plausible to regard the mind as the elemental unit of intellect (Déli, 2020a). This prompts us to explore how this mental unity can generate the astonishing complexity characterizing the social world, commencing our inquiry with the concept of spin.

The Cognitive Cycle Formulates Psychological Spin

Otto Stern and Walter Gerlach designed a test to pass silver atoms through an inhomogeneous magnetic field, now known as the Stern–Gerlach experiment. The anticipated outcome for particle distribution, mirroring classical spinning objects, was one of randomness and continuity. However, contrary to this expectation, the particles exhibited an even distribution between two distinct states: approximately half positioned in an upper state "spin up", and the remaining half in a lower state "spin down". This phenomenon pointed to spin as an inherent manifestation of angular momentum, constituting an essential elementary particle character.

According to information theory, the Carnot cycle reconstructs the past (liberates past information) through dissipative processes, and the reversed Carnot cycle belongs to the intelligent ones that control the future (Cox, 1979). The above rule might be prescient in cognitive science. Mistakes represent a failure to move forward (exothermic). In contrast, achievement, healthy behavior, and even retirement planning (Li et al., 2018) enhance mental energy (Déli et al., 2018) and inspire future focus (Deli, 2020b).

The concept of reversible perception cycles naturally aligns with a spin interpretation. A psychological grasp of spin prompts consideration of attitudes, which serve as instantaneous, context-dependent motivators for action. In high-pressure, high-arousal environments, situations laden with significant stakes, or moral dilemmas, attitudes wield particular influence over thought processes (Bechler et al., 2019).

Drawing a parallel between the half-integer spin characterizing fermions like electrons, which dictates atomic structure (only one electron per set of possible quantum numbers), and psychological spin, we find a potential link to the hierarchical structure of society (Wato et al., 2020). As such, attitudes are pivotal in shaping the quality of social relationships, career trajectories, and achievements.

To illustrate endothermic and exothermic psychological states, envision a soldier charging toward a formidable enemy (Roberts, 2019). In the time reversal, the warrior runs cowardly away from the enemy. However, a note of caution arises against interpreting time reversal in a purely literal sense. The orthogonal position signifies uncertain back-and-forth vacillations of the coward that hinder mental progress.

Subsequently, we explore how the reversible perception cycle gives rise to psychological spinor.

The Mind as a Spinor

"Adopting the right attitude can convert a negative stress into a positive one." —Hans Selye

Within the realm of spinors, undergoing a 360° phase shift results in a directional alteration of –1, a trait rooted in their inherent half-spin characteristics. This transformation effectively converts them into their opposite spin state (as illustrated in Figure 7). Spinors encapsulate the fermionic capacity to manifest up and down spin orientations. Fermions themselves are representative of spinors; their wave function necessitates a 720-degree rotation to return to the original state. In our conventional experiential context, spinors can be analogously depicted through objects like a Mobius band[35] (Figure 14).

As earlier chapters have exemplified, the mind readily succumbs to distractions and veers away from logic. The spinor character introduces an additional layer of unpredictability to this dynamic. A psychological spinor embodies the disposition-dependent direction of the perception cycle. Motivation becomes malleable under the influence of stress and folly, capable of morphing honesty into cynicism and humor into vexation. Pain disappears when serving a noble cause, but the absence of trust can potentially transmute a gentle caress into mistreatment and affectionate words into skepticism (Zhang et al., 2019). The spinor attributes of consciousness significantly contribute to psychology's replication (reproducibility) crisis. In this context, the spinor nature of the mind accentuates the pivotal role that attitude plays in shaping the trajectory of our life histories.

---

[35] A Möbius strip is a surface that can be formed by attaching the ends of a strip of paper together with a half-twist.

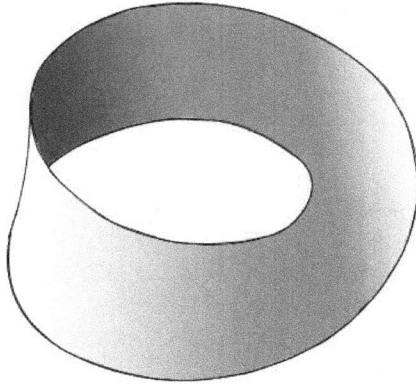

Figure 14. A spinor represented by the Mobius band. The particle and its environment represent the Mobius strip two sides. As the strip takes a complete turn, it turns inside out and upside down. It takes two full turns (720 degrees) to restore the starting condition.

The Down-spin Psychology: While occasional stress can heighten reward motivation (Ironside et al., 2018), persistently negative emotional states deplete motivation (Kool and Botvinick, 2018). The latter occurs because stress accumulates (refer to Figure 15). For example, gamma rhythms in the prefrontal lobe (Gao et al., 2022) or increased connectivity in resting-state networks (Gehrt et al., 2022; Xu et al., 2023) elevate emotional

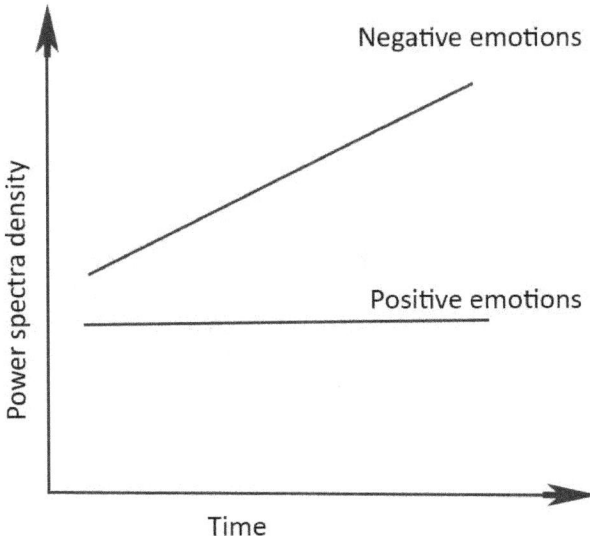

Figure 15. The cognitive burden of negative emotions. Negative emotions accumulate in the prefrontal cortex inducing high frequency gamma rhythms (drawn after Gao et al., 2022).

intensity (Sakai, 2020), leading to deterministic behaviors like impulsivity and problems in error correction (Aberg et al., 2023). Cognitive burdens negatively affect IQ (Gao et al., 2022), as illustrated in Figure 3.

Like all exothermic systems, the low-entropy brain discharges energy (Deli and Kisvarday, 2020; Deli et al., 2021) (refer to Table II). Energy dissipation can occur through vigilance (Manohar et al., 2018), criticism, destructive behavior, or violence. Long-term potentiation decreases resting complexity and entropy (Déli et al., 2021; Lin et al., 2022), propelling monotonous and repetitive thinking (Wang, 2020). Intellectual limitations are outward signs of the constrained degrees of freedom (Kaiser et al., 2016). A pessimistic fixation on past mistakes amplifies anger and sadness (Chen et al., 2023). Aggravation, violence, and regret represent exothermic energy release, compromising cognitive performance (Nuno-Perez et al., 2021). The above argument explains the early burnout of aggressive, restless, and conflict-driven personalities.

The wisdom of the German proverb, "Fear makes the wolf bigger than he is", highlights anxiety's power to impair one's resolve (Meeusen et al., 2020). Going through the motions without having a heart for the venture is no different from outright quitting. Corrupted decision-making (Lin et al., 2022) renders people susceptible to addiction, emotional challenges (Meijers et al., 2018), and cognitive impairments (Grieder et al., 2018).

Sigmund Freud has already pointed out the role of emotions in developing chronic conditions, such as cancer (Justman, 2020) and other adverse health outcomes (Ruan et al., 2020). The cognitive burden can suppress immune function (Cunnane et al., 2020), activate stress hormones (Alhussien and Dang, 2020), and cause mental problems (Grieder et al., 2018). The close link between psychological and physical health (reviewed by Koban et al., 2021) underlines the long-term health cost of aberrant thoughts and behavior.

The Up-spin Psychology: Given the significance of self-reliance in coping and adjustment, positive psychology underscores the role of a positive environment and mindset in fostering ambition and accomplishment (Schubert et al., 2019). There is energy in positive emotions and the willingness to act enthusiastically and decisively. The reversed Carnot cycle centers on a high resting entropy (Gao et al., 2022), which parallels temporal variability, signifying fluid intelligence (Wang, 2021). The greater degrees of freedom inspire optimism, openness (Zmigrod et al., 2019), and creativity (Shi et al., 2019).

The elevated entropy of the resting state signifies an accumulation of energy within the cycle, forming trust, belief, and confidence (Ryan and Deci, 2017). Mental energy is a structural facet of the brain that amplifies an individual's capacity to accomplish tasks through consistent cognitive exertion.

Like changing a vehicle's course involves a deliberate deceleration, rectifying errors necessitates a mental slowdown (Ryan and Deci, 2017). This mental deceleration, including practices like meditation, acceptance, and gratitude (Huang et al., 2020b), corresponds to a lower social temperature (O'Neill and Schoth, 2022; Déli et al., 2021). Positive states demand an initial cognitive investment; for instance, reward processing hinges on the production and release of thalamic neurotensin (Li et al., 2018). The question arises: how does this initial investment of energy translate into the accumulation of mental energy?

Landauer's principle has shown that erasing information from a memory storage device dissipates heat into the environment, increasing its entropy. According to Landauer, forgetting, which optimizes the growth and pruning of networks, is an intrinsic part of learning. In this context, learning potentially involves reciprocal changes in synapse connections (Dempsey et al., 2022), whereby the formation and retrieval of memories necessitate a concurrent loss of synapses. Despite the inclination to minimize uncertainty or entropy to streamline decision-making, pursuing certainty curtails the spectrum of possibilities, ultimately leading to an exothermic focus on the past.

In summary, positive emotions pinpoint exact moments while hiding their precise energy. In contrast, negative emotions make their potency evident while keeping their true origins elusive. The "law of orthogonality, acceptance" introduces further avenues for discussion.

The Quantum Zeno Effect: The quantum Zeno effect capitalizes on the particle-wave duality by employing frequent measurements centered around a specific configuration to halt the particle's temporal progression. An illustrative instance involves continuously observing an inherently unstable particle, effectively preventing its decay. Intriguingly, the quantum Zeno effect also finds its relevance within psychology. For example, conscious intentions can be viewed as a 'measurement' or 'observation'. Similar to the evolution of a quantum system, which can be slowed due to repeated measurements, frequently 'observing' or focusing on a specific intention, the associated brain dynamics might be prevented from changing, allowing the intention to have a sustained impact. Within quantum cognition models, extended evaluation or introspection stabilizes the initial viewpoint (Rossi, 2021).

Contextuality in Quantum Mechanics

"What we observe is not nature in itself but nature exposed to our method of questioning." — Werner Heisenberg

The world of quantum mechanics often challenges our everyday intuitions. One of the more perplexing elements is the idea of contextuality. In essence, the properties or behaviors of quantum systems are determined by the system and how we measure or observe them. When we talk about context in everyday life, we might think about the different circumstances or settings in which something happens. Similarly, in quantum mechanics, 'context' refers to the specific experimental setup or measurement choice. The outcome of a quantum event, like the spin of an electron, might depend on how and in what sequence we choose to measure it.

Classical systems, like a rolling ball or a swinging pendulum, behave the same way regardless of how we observe them. Their properties are intrinsic. But in the quantum world, the properties are not just inherent to the particle; they arise from the interplay between the particle and our chosen observation method.

Quantum Logic vs. Classical Logic: The peculiarity of quantum systems leads to the need for a new type of logic, known as quantum logic or nondistributive logic. Unlike classical logic, which obeys distributive laws (like distributive quality in algebra), quantum logic does not. In a departure from classical logic, in quantum mechanics, measurements can disturb the state of the system, and the sequence in which measurements are made can lead to different outcomes.

However, this 'logic' is not about ordering or ranking events. Instead, it's about understanding the hierarchy or relationship between different contexts (or experimental setups), particularly how they relate to or affect one another. The subsequent sections discuss complementarity,[36] the uncertainty principle, and wave-particle duality.

Context in Psychology

"To shift your life in a desired direction, you must powerfully shift your subconscious." — Kevin Michel

Perception is highly dependent on subjective variables as the interplay between system and context forms the very fabric of the quantum realm (Heelan, 1970). For instance, food may be highly desirable for a hungry

---

[36] Complementarity expresses the quality of requiring knowledge of both particle and wave aspects of a quantum system.

animal but repulsive for a sick one. Substantial incentives, whether bribes or threats, yield distorted, polarized, and thus unpredictable reactions (Kleiner, 2018).

In psychology, context refers to the role of circumstances in the outcome of a study. The mental state is malleable and sensitive to incentives and bribes. Context,[37] such as wording effects, can modify mood and response. Word recall can instantly conjure up related terms like 'ball', triggering softball, tennis ball, football, soccer ball, globe, ballplayer, and others. Social questionnaires and search engine optimization methods must consider these challenges.

The human ability to control Earth's resources is far superior to animals' simple existence, grazing or hunting for food. In Edwin Abbott's 1884 novella, *Flatland*, the citizens have a limited two-dimensional experience. However, the hero's experience of a three-dimensional world inspires him to imagine even higher spatial dimensions. Like the primitive Flatlander vision, we are bound by social and cultural beliefs.

Context dependence turns intelligence flexible, as opportunities can liberate a rainbow of talents, such as linguistic, logical, visual, spatial, musical, interpersonal, or intrapersonal skills. Overcoming personal and professional challenges can propel people to the top of their careers. Franklin Roosevelt's paralysis from polio did not keep him from becoming president, Sylvester Stallone's facial paralysis did not prevent him from acting, and Imre Kertész could write about survival in a concentration camp. These people became extraordinary by switching the 'context' of their lives.

Duality

One of the giants of quantum mechanics, Niels Bohr, introduced the notion of complementarity. It suggests that a single quantum entity (like an electron) can display dual behaviors (like particle and wave properties) but not simultaneously. Which behavior is observed depends on the experimental context.

The particle's ability to exhibit both point-like (discrete) and wave-like (probabilistic) behaviors is analogous to the metaphor of the blind men observing an elephant. The men must synthesize individual observations to comprehend the entirety of an elephant. Likewise, describing particles necessitates both particle-like and wave-like characterizations. The quantum quality of cognition (refer to Table I) is exemplified by phenomena such as the Necker cube (depicted in Figure 13). In the case of the Necker cube, the visual system manipulates ambiguous lines in a way

---

[37] Context is the circumstance forming of an event, statement, or idea, which is necessary for a complete understanding.

that only one of the possible three-dimensional interpretations is valid at any given time.

Surprisingly, the concept of wave-particle duality also applies to the way stimuli trigger our thoughts. When activated, the resting brain's wandering thoughts suddenly take on a distinct shape, like a wave transforming into a particle. Similarly, the relationship between experience and memory exhibits a similar switch between ambiguity and certainty, i.e., wave and particle. For instance, emotional affairs can alter our perception of time, and memories tend to accumulate based on their emotional significance. As a result, time is stretched and compressed to fit into our emotional story.

In the words of E.C.G. Sudarshan, "Conclusions, premises, feelings, and insights coexist in a manner that defies temporal order." The remarkable fluidity and vagueness of subconscious thoughts, emotions, and mental concepts contrast the discrete nature of convictions and faiths (Herzog et al., 2020). Despite gradual synaptic changes, beliefs are updated in discrete steps to build upon one another. This kind of cognitive duality leads us to uncertainty.

Uncertainty

"The mistake is thinking that there can be an antidote to the uncertainty."
— David Levithan

A speedometer provides us with the car's speed at any given moment. However, quantum mechanics makes pinning down a particle's location contingent on disrupting its momentum. The Heisenberg uncertainty principle forbids the simultaneous measurement of a quantum system's position and velocity. Uncertainty is a consequence of complementarity. The origin of this uncertainty within quantum mechanics is intriguing. Separating the wave function from the gravitational field conceals the quantum wave's evolution into a timeless dimension. When measured, an interaction occurs, altering quantum frequencies and the classical updating of the field curvature. The crux of uncertainty and wave-particle duality resides in the clear differentiation between the wave function and the field.

Uncertainty in Psychology: The brain's activations reflect the ebb and flow of probabilistic, reversible energy dynamics. Analogous to the above timelessness of the wave function, thoughts epitomize a state free from time constraints. The deeper the mental engagement, the more detached from time-keeping imagination becomes.

Decision-making ceases the brain's probabilistic computations, introducing certainty. Paradoxically, we often opt for unfavorable news over tormenting uncertainty. Although the search for a closing is deeply embedded in the mind, the value of information lies in its novelty for this predictive machinery.[38] Our engagement with adventure, horror, cliffhangers, and round-the-clock news channels stretches our sense of suspense like taffy candy. The heightened tension maintains our focus, offering a resolution just before reaching the point of no return.

The uncertainty principle also extends into intentional space and time within the human experience (Anderson et al., 2019). Heisenberg's principle implies between the time of the event and its momentum represented by motivation. In the case of positive experiences, interaction reveals their precise timing as they bubble forth with immediate enthusiasm. However, their holistic presence cannot offer quantification of their energy. There exists only complete happiness. Negative emotions, on the other hand, lead to a precise energy scaling of behaviors ranging from criticism and shouting to physical violence.

Nonetheless, their gradual accumulation in the brain's connection network is similar to sand piles. The slightest, most minor irritation can unleash their accumulated power onto unsuspecting recipients with the ferocity of a storm. Therefore, hurts create a lasting trace in the deep crevices of our minds, even if they do not generate a reply.

The Pauli Principle in Mental Interaction

"All things are made of atoms—little particles that move around in perpetual motion, attracting each other when they are a little distance apart, but repelling upon being squeezed into one another."
— Richard Feynman

Every alteration stems from the interaction of particles. Similar to minuscule gyroscopes, particles exhibit an inherent flow of energy known as spin,[39] which has the potential to impact the geometry of the spatial field. Matching geometries (spins) produce energy maxima (depicted in Figure 16A), which is unstable. Consequently, particles push each other to form contrasting spin configurations to fulfill the energy minima condition, an essential principle in physics (as illustrated in Figure 16B). This fundamental stipulation is referred to as the Pauli exclusion principle,[40] a central idea within the realm of quantum mechanics.

---

[38] New is the information's ability to update comprehension.

[39] Spin is always an active, up or down state, which reflects the wave function's connection with the gravity field.

[40] Pauli exclusion principle explains why two or more identical matter particles (same spin) cannot occupy the same quantum state.

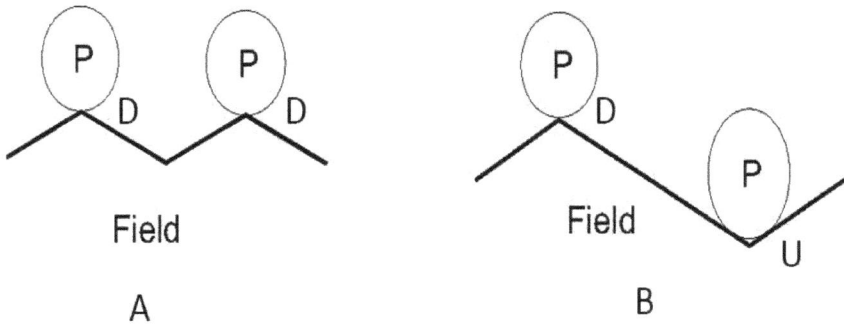

Figure 16. The geometric basis of the Pauli exclusion principle. "P" represents particles, while lines indicate field geometries. "D" and "U" represent down and up spin respectively. Identical spin particles form energy maxima, while opposite spin particles form a minimal energy formation.

Social Aspects of the Pauli Principle

Although our personal space is highly subjective and situation-dependent, relationships are characterized by precise emotional distance. The emotional distance also applies to attachments toward abstract entities or ideas. We automatically react to the slightest deviation in our drive to maintain emotional separation.[41] Detachment triggers a need for closeness, but neediness prompts the need for independence. The amygdala is responsible for the Pauli principle by controlling interpersonal distance in psychology.

Opposite Spin States: According to quantum mechanics, identical fermions cannot simultaneously occupy the same quantum state to neutralize spin momentums. The same rule is valid in psychology; close proximity to strangers or seeing emotionally disturbing images, even if unaware, causes us to feel aversion. Animals generally defend their personal space. However, a threat may trigger a fight or flight response[42] (Figure 10). In some cases, animals will freeze or play dead as a last resort to escape danger.

When interacting socially, opposition to others is spontaneous (as shown in Figure 16B). In small children, the discovery of their

---

[41] Emotional distance is the degree of trust toward others.

[42] An automatic physiological reaction to a stressful or frightening event activates the sympathetic nervous system and triggers an acute stress response that prepares the body to fight or flee.

independent selves is indicated by the profuse use of the word 'no'. Physics explains this contrarian disposition as the minimal energy configuration where aggravation represents down spin. However, it's important to note that the Pauli principle only applies to emotional proximity. Richard Feynman's quote also applies to society: people detest being squeezed into one another. Our strongest emotions are usually reserved for those we are strongly connected to, including the desire to take a break from their company. Unfortunately, this strong identification can also lead to historical conflicts between groups with shared backgrounds, such as the Sunni and Shia, Christians and Jews, Christians and Arabs, Jews and Muslims, and Democrats and Republicans.

Shared-down Spin: Mutual downward spin is the maximum energy formation of exothermic states, typified by shared antagonism and jealousy. At the neural level, reciprocal comparisons with others lead to rivalry (Peng and Xie, 2016; Saaty and Vargas, 2017) for the same things. However, rivalry injects similarities within the social classes while promoting social stratification. Ultimately, this hostile inclination results in a reluctance to offer compliments or ask for assistance. Therefore, shared downspins impose high cognitive costs and act against social cohesion.

Despite this phenomenon's substantial energy requirements and role in social discord, intense social stress, such as economic difficulties or worsening living conditions, sparks collective animosity and polarization. In numerous countries, the political landscape is characterized by the dominance of two major parties that champion opposing viewpoints on various issues. Competing companies often work against collaboration as they prioritize safeguarding and promoting their distinct systems.

On a global scale, countries organize into blocs with similar economic and political ideologies. This leads to polarization and prevents any force from dominating but, unfortunately, leads to significant wastage. The atmosphere of competition triggers activities like military buildup, espionage, and political maneuvering, ultimately exacerbating human suffering.

Shared-up Spin

Like shared-down spin, mutual up-spin is an energetically expensive formation. These configurations are exclusive to trusting environments or situations with significant personal distance, as discussed below.

A. Trusting Environments: Shared up spin requires social energy input in the form of patience and respect. Its existence is limited to trusting and altruistic social formations, such as family, friendship, school, or healthcare settings (as depicted in Figure 16A).

A, Mitigating the Pauli Principle via Distance: The philosopher Sextus once said, "Absence makes the heart grow fonder," which provides a social perspective to Richard Feynman's earlier quote. Distance, whether physical from our loved ones or social, from celebrities, politicians, and saints can lead to identification and attraction. The latter gives celebrities significant influence on shaping public opinion.

The Hierarchic Mind

In the words of Erwin Schrodinger, "Entanglement is not one but rather the characteristic trait of quantum mechanics." Entanglement facilitates the evolution of hidden energy disparities within a shared wave function. This correlation preserves the energy of the combined system, resulting in reciprocal energy changes for the constituent particles. Decoherence ends entanglement by separating the wave function into two distinct particles, revealing their individual attributes and positions. This process of entanglement, followed by interaction, gives rise to the vast intricacies of the cosmos.

Like the rich structure of space, social entanglement gives birth to hierarchical social evolution. Social entanglement engenders disparities in access to time (which can manifest as stress or contentment). Variations in mental comfort, particularly in terms of time perception, lead to the subordination of one participant to another. In societies marked by significant class divisions, this hierarchical structure helps suppress conflicts. However, with the advent of democratization, such class separation gives rise to thorny moral dilemmas. While entanglement plays a role in individual energy disparities, the stabilization of social stratification requires a gravity field (for a deeper exploration, refer to "The Structure of Society - Socioeconomic Status"). This dynamic gives rise to a duality between the properties of the wave function and the field.

Interference

> *"Nothing happens until something moves. When something vibrates, the electrons of the entire universe resonate with it. Everything is connected."* — Albert Einstein.

The Double-slit Experiment stands as the most renowned phenomenon within quantum mechanics. The name "double-slit" arises from a partition featuring two narrow apertures through which particles can pass. The wave function's peaks and troughs are superimposed within the experiment's context. This interplay of interference serves to amplify, diminish, or nullify the amplitudes, thereby generating a waveform marked by heightened or reduced amplitude (as depicted in Figure 17).

The Double-slit Experiment: The Double-slit Experiment involves the projection of photons (or other particles) onto a light-sensitive screen, which produces distinct dots. However, when these particles traverse a double-slit configuration, they create an interference pattern reminiscent of waves. The two setups highlight the concept of wave-particle duality: a quantum entity can exhibit characteristics of both a particle and a wave.

Amazingly, a measurement apparatus placed behind one of the slits abolishes interference. Even more astonishingly, the particle can land in a discrete position behind either of the slits on the screen. The conclusion is that the particle senses the device's presence even if traveling through the opposite slit. The perplexing question emerges: How does a measurement taken behind a single opening influence the outcomes of both possible paths? The answer finds its roots in wave-particle duality. Analogous to a computer's restart button function, a measurement reconfigures the whole particle's wave function. The exchange of energy with the field changes the field curvature and the particle energy function, subsequently canceling out interference.

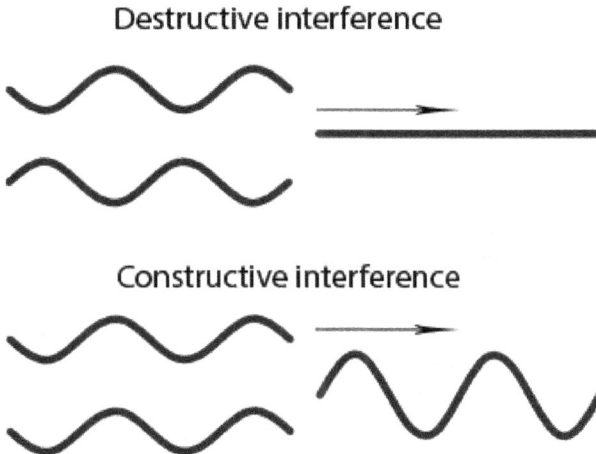

Destructive interference

Constructive interference

Figure 17. Interference between two same-frequency waves. Correlated waves (their peaks and troughs line up) generate interference. Destructive interference subtracts the waves' amplitudes, reducing their size (top). Constructive interference sums the amplitudes to create higher peaks (bottom).

Constructive and Destructive Interference

During interference, two correlated (coherent) waves[43] superimpose to form a resultant wave of greater or lesser amplitude. Its effects can be observed in all types of waves (for example, light, acoustic, and water waves). Constructive interference occurs when the crest of one wave meets the crest of another wave of the same frequency at the same point. Constructive interference sums the waves' amplitudes. In destructive interference, the crest of one wave meets the trough of another wave. In this case, the magnitude of the displacements is equal to the difference in the individual sizes. When the phases meet at intermediate points between these two extremes, the resulting shape of the summed waves lies between the minimum and maximum values.

Interference in Psychology

> *"It would be most satisfactory of all if physics and psyche could be seen as complementary aspects of the same reality."* — Wolfgang Pauli

Within psychology, repetition can align the similarities between separate events. In this context, interference encompasses aggregating or subtracting the structural peaks and troughs among experiences, thus influencing their impact. The manipulation of cognition through interference (as explored by Chandler, 1989) sheds light on the malleability of memories and knowledge. Often oblivious to the broader repercussions, individuals unfailingly form a temporal interference pattern in shopping habits, election results, or aversion (Xu and Schwarz, 2018).

Interference Theory: Interference theory, which posits that memory affects the pace of learning, has been a staple of psychology since 1892, predating the advent of quantum theory. Destructive interference hampers learning and memory performance, whereas constructive interference accelerates these aspects (as illustrated in Figure 17). The mental disturbance either complicates or simplifies retaining new information (Anderson, 2000). In the proactive scenario, existing knowledge overshadows new material. For instance, practicing sorting items with a different arrangement slows performance. In the retroactive case, recently acquired data, such as new phone numbers, can displace older ones from recall.

Destructive interference makes updating old information and modifying habits more challenging than acquiring fresh knowledge. However, interference can play a positive role when repeated exposure

---

[43] Correlated waves originate from the same source, or they have the same frequency.

unveils deeper meanings or shifts perspectives. Often, students attain a more mature comprehension of subjects as they progress to higher-level classes, and adult insight replaces a child's innocent outlook on events.

Social Interference

Operating instantaneously and independently of conscious participation, mental interference amplifies or dampens personal intentions. It can override economic rationality in various aspects of our social lives. For instance, aesthetically pleasing public spaces evoke a sense of security and encourage positive behavior. The aroma of household cleaners fosters cleanliness and activities that promote prosocial behavior inspire charity. Disorderly surroundings, such as random piles of trash, can stimulate feelings of stress, asocial thoughts, and phobias, leading to stereotyping, carelessness, and delinquent conduct. Similarly, although playing violent games can alleviate stress, its long-term antisocial effects can lead to aggressive behavior. Interference gives rise to social phenomena through temporal waves and bursts.

Impact on Decision-making: Temporal interference wields significant influence over decision-making processes. For instance, in a game reported by Savage (1954), where participants have an equal chance of winning $200 or losing $100, approximately 69 percent of winners and 59% of losers choose to play another round. Logically, not knowing the outcome should result in an average between the winners' and losers' percentages (64%). Surprisingly, uncertainty prompts interference, leading to a drop in the players' combined ratio to 36 percent (Tversky and Shafir, 1992). Without feedback, the mind exists in a state of quantum uncertainty. Gaining knowledge of the outcome acts as a measurement device, which collapses the wave function and forms interaction. Interaction opens the path to a logical decision.

Prisoner's Dilemma: The prisoner's dilemma is a renowned psychological analogy to the double-slit paradox (Stewart and Plotkin, 2013). In this scenario, two separately incarcerated criminals face prison sentences. They can either collaborate with their partner (by remaining silent) or betray each other (by confessing). Cooperation results in a reduced punishment, while defection offers freedom but imposes a significantly longer sentence on the cooperative companion. Remarkably, individuals overwhelmingly opt to support and cooperate with each other, an inclination best explained by the interference theory. Without interaction, the mind is in quantum limbo and subject to interference.

Interference can complicate problem-solving. Uncertainty surrounding event outcomes can hinder progress in one's life. Again, the absence of closure inhibits decision-making. The mental ambiguity can maintain persistent efforts to search for a lost treasure, a past love, or a scientific breakthrough. The process may appear as loyalty, determination, curiosity, or laziness for outside observers. Adopting an external perspective on the problem is a form of interaction that resolves uncertainty.

Investment Bubbles: Advertisements often portray idealized dreams that exaggerate appeal through interference with natural desires. Significant and widely publicized financial gains pique interest, leading to the formation of investment bubbles. When the rapid price growth halts, enthusiasm turns sour, resulting in destructive interference. This shift prompts avoidance behavior, triggering a frenzied sell-off that causes the price to plummet. These losses offset the previously unsubstantiated gains. A parallel instance is the bandwagon effect, explored further in the section "Social Lasing".

Social Lasing

> *"Realize that everything connects to everything else."* — Leonardo Da Vinci

The process of laser amplification involves the stimulation of an atom with the help of multiplier mirrors to emit identical photons, resulting in the production of coherent light. This type of light can be focused on a small area or transmitted over long distances without diverging, making it useful for various applications such as communication, pointing, and cutting/marking materials.

Social Lasing: Social interference gives rise to investment bubbles, but social lasing is an even more effective collective motivator. Social laser intensifies social energy[44] into coherent actions. On social media, reposting boosts the amplitude and coherence of information waves (social energy), serving as a multiplier mirror (Khrennikov, 2023).

Amplified discontent can trigger revolutions and protest movements like Brexit or the Arab Spring. Therefore, social lasers transform information radiation into specified behavior and physical action sequences, confirming that the brain operates as a thermodynamic machine.

---

[44] Social energy is a special form of motivation boosting decision-making.

The Fermionic Mind Hypothesis (FMH)

"We are all agreed that your theory is crazy. The question that divides us is whether it is crazy enough to have a chance of being correct."
— Niels Bohr

Imre Lakatos, the Hungarian-born science philosopher, pointed out that some scientific theories may not be directly falsified. A hypothesis should be assessed by its capacity to generate novel insights and predictions of previously unknown phenomena (Doerig et al., 2022).

Consciousness embodies a self-regulating system akin to a particle, specifically a fermion (as outlined in Table I). The mind can uphold its internal stability and enduring nature while seamlessly adapting to the ever-evolving surroundings. Put simply, despite constant fluctuations, the mind remains fundamentally unchanging—an attribute that mirrors the electron,[45] the most familiar fermion.

The Holographic Mind: The Fermionic Mind Hypothesis (FMH) stands out as a distinct theory by elucidating consciousness attributes through reference to the particle structure in physics. The continuous flow of information from the environment projects an inverted representation onto the cortical surface in an involuntary process—a phenomenon aligned with the second law of thermodynamics. The mind, guided by an inherent comprehension of the environment and its governing laws, is an innate generator of remarkably unified and holographic consciousness.

Particle-wave Duality: Despite the necessity for diligent practice, acquiring skills like walking or riding a bike is an abrupt accomplishment, with the skill emerging in a sudden flash of clarity. The same vague uncertainty characterizes the thinking process or the initial phase of the perception cycle. Nevertheless, decision-making or recognition crystalizes the chaos in a discrete energy switch that triggers cognitive clarity. The probabilistic nature of the subconscious and the field-like stability of self-consciousness (Wolff et al., 2019) are reminiscent of the wave-particle duality. Surprisingly, the reversal of the process also represents a wave-particle transformation. Activation by stimulus reduces the diffuse exploratory resting thoughts into sensory action. Therefore, our experience can only offer us the fluidity of the wave function intercepted by flashes of recognition.

This duality leads to one of the psyche's most perplexing, counterintuitive, and challenging aspects. The introspectively accessible, phenomenal facets of the mind lie at the core of the mind-body problem. Instead of attempting to deconstruct this complex puzzle step by step, a resolution can be found by recognizing the mind as the most fundamental unit of intelligence.

The Observer Mind: Evolution didn't introduce entirely new laws for life and consciousness; instead, the brain adapts to its environment by incorporating the regulatory principles of the laws of physics. Consequently, intellect can be defined as an endothermic process that increases complexity. Consciousness signifies an awareness of the mind's distinct observer status over the self-regulating brain. Our beliefs, internal dialogues, and stable sense of self originate from this resting state. Its particle-like isolation bestows the mind with its self-referential characteristic—an individual's internal, subjective perspective.

Temporal Continuity: As perception revolves around the resting state, it shows reversibility and probabilistic nature. Moreover, the comprehension of stimulus triggers thoughts and emotions, which reach back to the regulatory foundations of physics for conceptual orientation in time. Therefore, perception is a thermodynamic cycle that confers attitudes with temporal direction and motivational energy. Temporal directionality represents cognitive spin. Up spin comes from an endothermic process, while down spin is exothermic. As particle spin entails intrinsic angular momentum, psychological spin corresponds to intrinsic motivation. Consequently, the mathematical tools developed for addressing quantum mechanics problems are highly applicable to addressing specific queries (and paradoxes) in psychology and the social sciences.

We have observed that the generation of energy within the cognitive cycle depends on the significance of stimuli, and the scope of free will is confined to determining the direction of that cycle. Notably, abrupt shifts in the neural dynamics form between the hippocampus and prefrontal regions, stemming from bifurcations (Sridhar et al., 2021). These contrasting gradients, representing past or future behavioral trajectories, support the above conclusions.

Dilation of Time Perception: In analogy to the phenomenon of time dilation caused by photons, the motivational potency of emotions must lie in their ability to expand the perception of time. Emotions like hunger and fear motivate engagement with the external world by altering time perception to fulfill the thermodynamic imperative of restoring the resting state. Hence, emotions are the fundamental drivers of motivation (Déli, 2020a, b); their remarkable capacity to generate action gives rise to the belief in free will. Moreover, the thermodynamic analysis of perception makes it clear that quantum phenomena can ensure energy conservation. The energy conservation of entangled particle evolution is only possible if they are part of one combined system.

The Temporal Fermion: The origin of mental regulation can be traced back to general relativity and quantum mechanics, reinforcing the fundamental coherence between the mind and the physical realm. Just as fermions adhere to the stationary action principle in physics, the mind's predictive processes epitomize the most economical mental path toward the future. Supersymmetry theory posits the existence of a super-partner for each fermion. By extension, it's conceivable that the mind is the temporal super-partner to fermions.

FMH demonstrates that evolution can harness physical laws to motivate self-preservation and survival. As a result, the universality of physical laws extends to the realm of intellect. Physics emerges as the exclusive guiding principle behind the intricately harmonious cosmos. Rather than devising novel forces and laws, biological evolution integrated the existing physical laws into the brain's framework to forge the most efficient regulatory mechanism—the intelligent mind.

Consequences of the Fermionic Mind Hypothesis: Drawing inspiration from Imre Lakatos' ideas, the Fermionic Mind Hypothesis (FMH) unveils fresh explanations and insights into the brain's inner workings, addressing pivotal queries within the field of consciousness science: (1) The basis of the brain's self-regulation is the resting balance. (2) The evoked cycle's search for thermodynamic balance occurs through emotions. Indeed, residual emotional charge corrupts the resting coherence and openness (Gehrt et al., 2022). (3) The neural bifurcations between the hippocampus and prefrontal regions (Sridhar et al., 2021) embody contrasting gradients, signifying the divergence between past and future focal paths—a foundational element of quantum probability. (4) Emotions' exceptional motivational power to incite action occurs by slowing time perception. The temporal orientation of the mind vis-à-vis material systems represents orthogonality, turning the dilations of time into the dilation of time perception. Therefore, time perception represents field curvature. (5) Decision-making transforms probabilistic thoughts and concepts into the stable convictions that characterize the resting state.

The brain's cognitive cycle can run in reverse, making perception uncertain. This reversal provides a spin-like quality due to the opposite energy direction. The switch between the irreversible resting state and the reversible cognitive cycle is similar to the wave-particle duality. As a result, the cerebral system adapts to quantum characteristics.

In physics, quantum mechanics relates to the behavior of matter and energy on the scales of atoms and subatomic particles. How can the brain, a classical system, give rise to quantum cognition? The answer comes from the brain's operational dimension. Perception is an orthogonal transformation of the spatial sensory signals into a temporal rhythm, the fundamental operating principle of the mind. While elementary particles are regulated by gravitational curvature (contraction or expansion of space), cognition is sensitive to the perception of time, a quality transmitted by emotions. Perception turns time into a flexible field, enabling quantum computation. The perpendicular transformation from space to time endows cognition with quantum computing abilities. However, quantum cognition is a temporal quantum computer.

Comprehending physical laws fundamentally shapes our intellect— the understanding of gravity, for instance, empowers us to achieve balance, engage in running, and propel a ball through the air. As these physical laws are assimilated, the mind takes on a particle-like organization.

As we reflect on the universe we are part of, our minds assume the form of a distinctive, particle-like cosmos. The particle-like stability ensures the continuum of conscious selfhood from infancy to old age, transcending the profound changes occurring to the body and brain. Consequently, the essential nature of an individual's mind reflects the entirety of the universe, corroborating the Eastern concept of the unity of existence—Atman and Brahman are one.

To comprehend particles is to comprehend the mind and, by extension, to understand the universe itself.

Hard-to-Explain Experiences

Our brains are constantly trying to predict the future, which can sometimes lead to errors in perception, such as sensory hallucination of expected objects that are not there. Pareidolia is the mind's tendency to find meaningful patterns, especially faces, in random data. Making up things is even easier in dreams. Dreams can be a wild and bizarre carnival of strange and unconventional narratives created by our brains by combining different memories and experiences. However, unlike our wakeful cognitive processes, we rarely question the absurdities we encounter in our dreams.

Our brains can be selective about the information we process, leading to inattentional blindness. This means we can miss apparent things in our visual field because our attention is elsewhere. This emphasizes the limits of our perception and how our brains prioritize information. The strange feeling that a new situation has occurred before is known as *déjà vu*, and it continues to puzzle scientists.

The above occurrences have no current scientific explanation. Still, the brain's autonomic self-regulation can provide a plausible interpretation. Sensory perception is driven unceasingly by memories and experiences into a projected anticipation. Pareidolia and dreams have similar origins; they can be kaleidoscopes of our fears, experiences, expectations, or discoveries. A great story is of the nineteenth-century German chemist August Kekulé, who claimed to have pictured the ring structure of benzene[46] after dreaming of a snake eating its own tail.

Finally, neuroscience and psychology have tried to explain many human characteristics due to millions of years of evolutionary selection pressures. However, the temporal particle offers a radically different paradigm that defies simple evolutionary explanation. Although the mental particle is subjected to evolutionary pressure, its fundamental character traits emerge from thermodynamic pressures rather than predator-prey dynamics.

Future Directions

We are losing control over our future due to climate change and economic inequality, which increasingly force migration and human suffering. Additionally, we face ongoing conflict and war. To address these issues, we must prioritize integrity over individuality, empathy over rivalry, solidarity over patriotism, wisdom over white-collar superiority, and consensus over discipline. According to traditional psychology, these are intractable issues. However, the physical approach to consciousness can provide a solid path toward a transformation, which is organic, originating at the individual level, and can act in synergy with leadership to radically transform our social fabric.

The preceding discussions have centered on gravity and electromagnetism—two fundamental forces shaping our everyday experiences. In contrast, nuclear forces operate within the atomic range. Among these forces, the weak interaction is responsible for the breakdown of unstable elements, and the strong interaction, also known as the strong nuclear force, is responsible for confining quarks within particles called hadrons. This force boasts two distinct ranges, operating through corresponding force carriers. The weaker of the two binds protons and neutrons (nucleons) within the atomic nucleus. The stronger one is the gluon, which ensures the cohesion of quarks within hadrons. The impossibility of quark separation illustrates the gluons' immense strength. When quarks are separated by a distance surpassing the diameter of an

---

[46] An aromatic functional group characterized by a ring of six carbon atoms, bonded by alternating single and double bonds.

atomic nucleus, the result is the emergence of additional quark-antiquark pairs, which occupy the intervening space.

These forces' social analogies are speculative. The social equivalent of strong nuclear force is trust, an emotion regulating emotional closeness. Trust takes root in shared dignity and fosters a sense of unity, bridging differences among people and even animals. Trust invokes respect while liberating from fear. Similar to the behavior of the strong nuclear force, its hold diminishes when in close quarters but gains potency as the distance between individuals grows. Trust endures through loss by reformulating itself in the community. Dignity fosters an environment of trust, exerting its influence on the broader society.

The following parallel processes will fundamentally transform our social structure. The inevitable introduction of universal basic income will spare people from losing their moral compass due to necessity, as explained by Festinger's cognitive dissonance theory. The slowing and reversal of population growth combined with increasing productivity and creative potential of the individual due to technological development will dramatically increase the value and dignity of the individual. This will sweep away the current, highly competitive, often cutthroat business environment, where resorting to misleading and cheating customers is the norm. The employees are often subjected to abusive treatment in the interest of profit.

As the social environment changes, so does the social structure. People can be compassionate about events affecting the furthest corner of the Earth. Therefore, the interest of others is just as important as that of the self.

As demonstrated, the same phenomena can adopt distinct interpretations when viewed through classical and quantum physics lenses. Although science strives for a unique explanation for various phenomena, in reality, multiple insights often can complement each other. The evolution of scientific thought throughout history supports the latter argument. During the history of science, explanations of various phenomena have grown in complexity as science has delved deeper into details and achieved a more profound comprehension. In fact, diverse perspectives can yield more intuitive insights.

Chapter 5

Behavior Regulation

"Optimism is the one quality more associated with success and happiness than any other."

— Brian Tracy

In the preceding sections, we observed that animal survival hinges on the astute interpretation of the external environment, which serves the maintenance of biological functioning and physiological and psychological balance. Emotions prompt the actions to achieve those aims. Just as energy transitions between various forms, emotions transform and motivate behaviors. Emotional motivation transforms into work representing muscle movement. Drawing a parallel to photons, which influence electron behavior, sentiments generate intentions and behavior. Just as photons represent elemental forces, emotions constitute fundamental drivers of motivation. This chapter discusses the role of emotions in normal human physiology.

Intelligence

"The measure of intelligence is the ability to change." — Albert Einstein

At its core, intelligence is an adaptive mechanism that allows individuals to interact optimally with their environment. As remembered, the spider's intelligence rests in its ability to catch prey by sensing the net vibrations. While animals primarily rely on genetic adaptations over long periods, humans have the unique capability to create tools, technology, and social constructs that shape their environment. Human ingenuity and insight led to social evolution, which increasingly transforms culture and social organization. In turn, technological development increases the productivity and creative potential of the individual.

Forms of Intelligence: Human intelligence is multifaceted. Gardner's theory expands the traditional view of intelligence to include musical, bodily-kinesthetic, and interpersonal intelligence. These various forms confirm the perspective of parents and educators that students think and learn in many different ways. These diverse forms of intelligence emphasize that human potential is not strictly tied to logical or mathematical abilities.

Intelligence is highly heritable and predicts important educational, occupational, and health outcomes better than any other trait (Plomin and von Stumm, 2018). Nevertheless, it is also malleable; for example, major systematic interventions, such as education (Brinch and Galloway, 2012), diet (Protzko, 2016), or adoption (Duyme, 1999) can make significant differences in measures of intelligence in children.

Mental Energy: Nevertheless, significant achievements require a serious, long-term commitment. Synaptic adaptability ensures mental flexibility, while traits such as self-confidence, a proactive mindset (Yang et al., 2020), and goal-driven orientation permit a substantial concentration and focus (Shimazaki, 2020) to overcome external and internal barriers. The general factor of intelligence, the g factor, was discovered by Spearman in 1904. He noticed that intelligence is a complex trait penetrating many behavioral and psychological outcomes, including educational attainment, occupational status, health, and longevity. Therefore, increasingly, several aspects of intelligence are considered.

Intelligence Quotient (IQ) measures cognitive comprehension and problem-solving skills. Emotional Quotient (EQ) represents emotional management and self-awareness. Social skills and interaction capabilities are quantified by Social Quotient (SQ). Finally, the Adversity Quotient (AQ) concerns resilience and overcoming adversity. While IQ measures academic abilities, EQ is often a better predictor of life and work success.

Emotional Intelligence and Energy: Analogous to potential energy in physics, emotional intelligence requires consistent cultivation (Kong et al., 2019) (Figure 7). While rest can relieve fatigue (MacCann et al., 2019), a lack of mental energy impedes accomplishments (Meeusen et al., 2020). Motivation, perseverance, and well-being are crucial in determining outcomes in various fields, from professional achievements to athletic performance (Nicolas et al., 2019).

Thermodynamics and Success: The brain's thermodynamic investigation unveils a reality bereft of shortcuts to success. While we often attribute achievement to talent, substantial accomplishments are born from persistent and purpose-driven endeavors, mental rigor (Inzlicht et al., 2018), unyielding perseverance, and an unwavering commitment to hard work (Figure 6). Various practices, such as meditation, meaning, purpose, and acceptance, can help improve well-being and cope with personal and professional challenges (Chang et al., 2019).

Supportive Environments: The intricate energy balance of the human psyche underscores the importance of a positive social and educational environment (Crawford et al., 2020). Such environments foster growth and offer the emotional stability necessary for individuals to remain committed to their pursuits over extended periods. Even when progress appears incremental, an endothermic process provides emotional balance, resilience, and the tenacity needed for sustained, long-term commitment (as illustrated in Figure 7).

Within a dynamic environment, biological systems exist as endothermic entities, relying on continuous energy input to facilitate growth and adaptation. Since mental evolution thrives on adaptability and change, novelty becomes a fundamental prerequisite for intellectual development.

Emotions in Well-being

"There is no path to happiness: happiness is the path." — Buddha

Traditionally, intellect was considered the domain of mathematical logic, the faculty of reasoning and understanding objectively and abstractly. Increasingly, the role of emotions in intelligence is beginning to be recognized. Contemporary research also substantiates ancient insights that optimism, trust, and confidence are integral to psychological well-being. Sages, philosophers, and artists over millennia preceded scientists in recognizing the health advantages associated with positive emotions.

The Role of Irreversibility: In physical systems, irreversibility drives the movement toward a state of high entropy equilibrium, a concept known as the arrow of time. The brain's resting activations' remarkable irreversibility (Kolvoort et al., 2020) optimizes the action repertoire between past and future (Déli et al., 2022). However, its relevance contrasts sharply with the arrow of time, which degrades the system's degrees of freedom. Although the irreversibility of the resting state represents a state of mental equilibrium, it is an active state with future directionality. It enhances the scope of actions toward increased freedom (Zanin et al., 2020), creativity, and novelty (Jeffery and Rovelli, 2020).

The accumulation of negative influences taints the mind with inflexibility and generates past focus. Long-term stress lowers resting entropy and gives rise to reversible activations (Kolvoort et al., 2020). These reversible resting activations, which show time symmetry and monotony, are also characterized by pathological conditions as evidenced in major depression (Stringer et al., 2019; Wang, 2020). There are no quick remedies for diminished mental energy, which often introduces rigidity and pessimistic past focus. The mind tied to the past is destined for a reduced quality of life (Sizemore et al., 2018).

Applying the principles of thermodynamics to the domain of perception elucidates these contrasts, enabling the analysis of emotions through the lens of physics.

Positive Psychology

The contemporary interpretation of karma has paved the way for positive psychology, a field dedicated to investigating the advantages of optimistic thinking and improving the quality of life. It is defined by Martin Seligman and Mihaly Csikszentmihalyi as "the scientific study of positive human functioning and flourishing on multiple levels that include the biological, personal, relational, institutional, cultural, and global dimensions of life". It offers evidence-based interventions to boost well-being and enhance life satisfaction by exploring concepts like flow, gratitude, resilience, and hope. It provides tools for individuals to lead more fulfilling lives and for societies to foster more thriving citizens.

The Minnesota Nun Study delves into the life trajectories of nuns, revealing that complexity, fluency, and imagination in youth protect against later cognitive impairments like Alzheimer's (Iacono et al., 2009). Positive thought processes, a sense of dignity, and forward-looking anticipation foster well-being and healing and can contribute to longevity. Traits like faith, love, and courage catalyze enduring enthusiasm, generosity, and cooperation. Even in broken relationships or loss, one can find transcendence and closure through emotional healing.

In today's rapidly evolving society, psychological well-being is intricately linked to personal growth and expanding possibilities (Kolvoort et al., 2020). While happiness serves as a motivator for striving toward a brighter future, the mere act of feigning happiness proves ineffective. Deceiving our employers, friends, or partners might be possible, but deceiving our own minds remains an impossible challenge. The principles of positive psychology extend to communities as well. Physical proximity, touch, and embracing signify group unity, cohesion, and mutual trust.

The Hazards of Superficial Joy: Superficial emotions like egocentricity and boastfulness can be likened to empty calories that sow seeds of insecurity and hollowness. The grand promises of overconfident young talents might transform into disillusionment and bitterness, leading to lives bereft of meaning. Comparable to a mirage, the allure of luxury and comfort distorts our perspectives, continuously altering our expectations.

Emotional dynamics, akin to electromagnetism, highlight the relative nature of comparisons; hedonic adaptation erodes the joy stemming from superficial accomplishments. Substantial windfalls, such as promotions or lottery wins, may offer only fleeting contentment. Sustained well-being necessitates facing and overcoming challenges.

Love: Love, an affection that emanates warmth and intimacy, invokes a mental slowing down, relaxing and elevating the soul. Beauty acts as the spark for both romantic and parental love. In the eyes of a lover, beauty becomes the sole focus, creating an intensifying spiral of affection that perpetuates itself through attraction and appreciation. The recognition of beauty is not limited to humans; animal courtship validates physical perfection and cognitive prowess through graceful and harmonious movements, often resembling a dance. The process also includes securing and creating places for mating and nurturing offspring.

Social well-being is built on personal dignity, engendering trust, and social mobility, the subject of the next section. Significant negative correlations exist between stress and the components of emotional intelligence, such as emotional awareness and expression, emotional thinking, and emotional regulation (Jung et al., 2019). High levels of anger, a component of stress, were significantly related to poor emotional regulation. Perhaps the most notable ancient understanding of the long-term health and career consequences of emotional states is the Eastern concept of karma.

Karma

> *"As she has planted, so does she harvest; such is the field of karma."*
> — Sri Guru Granth Sahib

The philosophy of karma holds a significant connection with the concept of rebirth in Indian religions. However, a broader interpretation of karma, derived from the Sanskrit term "Karman", encompasses action, effect, or fate. It entails the belief that our present actions somehow shape our future outcomes. Acts driven by good intentions and virtuous deeds are believed to contribute to future happiness. In contrast, malevolent intentions and immoral actions are thought to lead to suffering. The latter perspective is shared across various religions and philosophical systems, exploring the interplay between reciprocal action outcomes in the 'Just-world' hypothesis.

The Difference from Lenz's Law: Lenz's law acts instantaneously to oppose changes in the external environment. Instead, karma represents the additive nature of our efforts by compounding mental power and its external consequences. Moreover, karma reflects long-term field effects, immune to human evaluations of time, as described by Benjamin Bayani. Therefore, karma is the consequence of general relativity operating in the social sphere. Our daily actions act on the curvature of the social field to bring long-term changes in personal fortunes.

Supportive surroundings bolster self-assurance, while stress hampers focus and curtails performance (Nuno-Perez et al., 2021). This intertwined relationship leads to karmic consequences. According to karma, the power of one's attitude (the energy emitted) holds greater significance than circumstantial factors. Confidence draws in support, whereas harassment invites mistreatment.

Bad Karma

Pessimistic attitudes narrow focus (Izaki and Ogawa, 2023), limiting mental freedom (Figure 18). A lack of self-esteem denies the capacity to cultivate genuine relationships, fostering selfishness and emotional instability. These emotional swings create a self-perpetuating cycle that drives one toward unhappiness and accelerates cognitive aging.

The Perils of Deceit: Cheating corrodes the soul, breeding distrust and vigilance. Once doubt infiltrates the mind, it permeates every facet of judgment, curtails possibilities, and complicates life. Lenz's law shows that a deceitful person projects onto others his own skepticism. A careless carpenter anticipates the collapse of houses, and a neglectful tailor or shoemaker is critical of the goods others make.

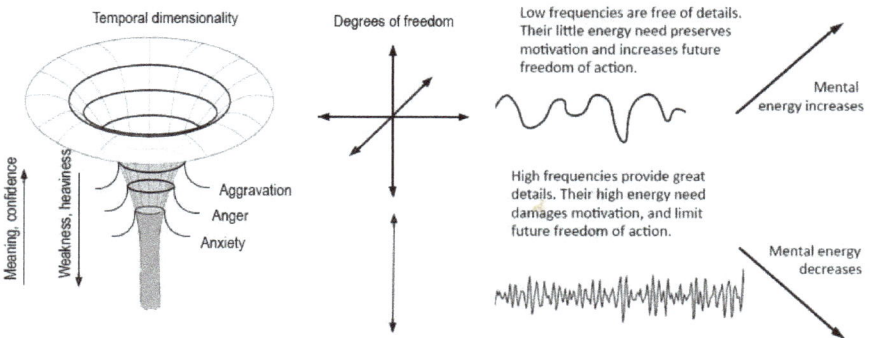

Figure 18. The effects of emotions on mental freedom. Positive emotions and meaning expand freedom (top). On the other hand, the higher energy needs of aggravation, anger, and anxiety waste time and energy, which decrease the future freedom of action. (Figure is courtesy of Deli, 2020a.)

Good Karma

A positive attitude and proactive actions cultivate confidence and an optimistic perspective, forming the essential components for achieving success (as depicted in Figure 19). This positive approach yields a dual effect: firstly, it garners the respect of peers, clients, and our social circle; secondly, it nurtures self-assurance and contentment. Dependable and conscientious individuals foster their personal achievements and inspire success in others. The significance of ethical conduct endures even if no one notices it. Therefore, they can afford not to respond to provocation while courageously confronting injustice when necessary. Even personal tragedies, such as the family tragedy, spur positive responses. For example, they might advocate for regulatory and social reforms.

Optimism: Optimism, a hopeful outlook on the favorable outcome of endeavors, empowers confident individuals. In the face of uncertainties, setbacks do not dampen their spirits for long. Armed with resolve, they transform ambiguous situations into triumphant narratives (Treadway et al., 2012). Consequently, a discerning mind neither endures nor carries the burden of maltreatment or injury over the long term. Being unburdened by chronic stress contributes to a healthier and longer life span.

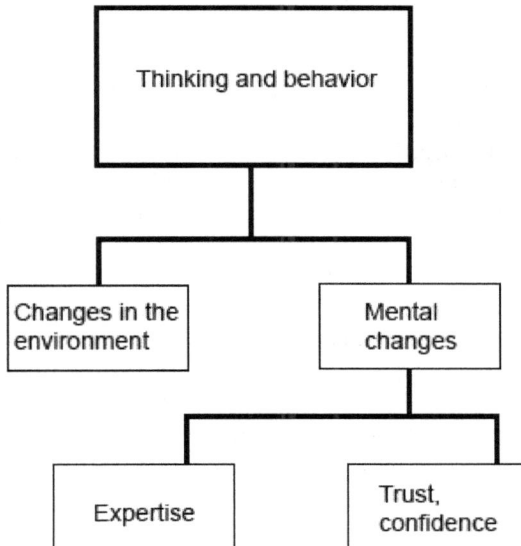

Figure 19. The origins of positive karma. Response to a stimulus modifies both the brain (mind) and the surroundings. Likewise, learning produces expert knowledge and confidence, a compelling combination for success.

The mantra, "I think I can, I think I can," echoes in the Little Engine's and countless others' sentiments (Piper, 1930). Self-efficacy, characterized by the belief in one's ability to execute tasks, is critical to finding solutions. Emotional stability facilitates the commitment to self-improvement, inspiring the same in others. They are healthier, happier, and wealthier and contribute to a well-functioning society (Moffitt et al., 2011). Through continuous interaction with the social milieu, confident people maintain a youthful, active, and lively existence even in old age.

Understanding karma liberates us from the grasp of vengeance and remorse. It frees us from expecting instant results from our efforts.

Social Mobility

"A human must turn information into intelligence or knowledge."
— Grace Hopper

Emotional well-being is fundamental for individual flourishing and reaching one's potential. This is also connected to the ability to contribute to society. The wide array of human skills, encompassing domains like music, language, organizational aptitude, and spatial thinking (stereopsis), combined with universal education and democratic institutions, are supposed to provide everyone with equal opportunities for achievement.

The adaptability of synaptic connections in response to changing circumstances suggests that mental agility could be crucial in intellectual prowess and navigating challenges through effective problem-solving. The individual notion of intelligence is cognitive flexibility (Birney and Beckmann, 2022). Cognitive flexibility enables us to switch between concepts or adapt behavior to achieve goals in a novel or changing environment. Why does social mobility seem out of reach for those at the lowest rungs of the economic ladder?

Interestingly, an individual's personal approach to education often wields more influence than their inherent abilities. Recent studies highlight that intellectual deprivation, characterized by a scarcity mentality and a sense of neediness, emerges as a primary factor behind untapped potential (Kearney and Levine, 2016). A mindset of disadvantage has been linked to a 13-point decrease in IQ, akin to the cognitive effects of sleep deprivation. The absence of financial security or overreliance on financial concerns or social status can stifle creativity (Mani et al., 2013), a topic explored in the subsequent section.

In conclusion, without a basic sense of security, democracy remains illusory.

Creativity

"Creativity takes courage." — Henri Matisse

Every passing generation must grapple with technological, cultural, and religious shifts. Still, our technologically driven society inspires a fixation on advancement. Creativity is steadily becoming an essential asset in our multifaceted lives through its role in inventing novel technologies, conquering ailments, and unraveling societal dilemmas. Nevertheless, research demonstrates that progressive technological environments nurture intelligence without necessarily impacting creativity (Makharia et al., 2016).

The epiphanic moments of inventors, artists, and scientists are celebrated across the annals of human history. Nevertheless, creativity is not the privilege of the few. Children's boundless imagination is a testament to the inherent human capacity for creativity. How can we ensure that this abundant gift endures into adulthood? How do we cultivate and nurture creativity?

While necessity might be the catalyst for invention, a positive mindset is the conduit for giving birth to something new. Innovation, which addresses practical problems, may not be a pressing concern for the affluent. Conversely, individuals trapped in poverty are often unable to rise above their hopelessness. The elixir of success demands a blend of stress and an ample measure of ambition mixed with purposeful work. A purpose diminishes anxiety, mitigates conflicts, and even modulates pain sensitivity (Hooker et al., 2017). Working for a goal expedites problem-solving for humans, animals, or even robots. For example, programming a task into robotic simulations yields marked improvements in problem-solving prowess (Wissner-Gross and Freer, 2013).

Creativity as a Process

Engaging in empty entertainment is draining. Picasso's assertion, "When I work, I relax," underscores the delight and contentment inherent in creativity, a state often referred to as "flow".[47] Flow embodies a complete immersion of the mind, fostering intense concentration and unwavering engagement. The actor becomes one with their role, the musician harmonizes with their music, and the painter channels their imagination into tangible form. Achieving this harmonious resonance of the soul

---

[47] Csikszentmihalyi introduced *flow*, a state of focused abandon, in the seventies (Csikszentmihalyi, 1980).

necessitates assiduous technical groundwork to channel intuition. Thus, creativity emerges as an authentic and ego-transcending alignment of the soul.

The Chaos of Creativity: Astrophysicist de Grasse Tyson succinctly notes, "Rational thoughts never drive people's creativity as the way emotions do." Analytical thinking demands focused attention, whereas creativity requires a dynamic mental landscape to navigate the perplexing, challenging, and sometimes overwhelming phase of problem-solving. In this chaotic process, the brain's innate regulation compresses, processes, and sieves through conflicting information. While stress amplifies the perception of time and can lead to impatience, a receptive and relaxed mind can navigate mental turmoil. Creativity defies planning; it arises unexpectedly when the focus broadens to encompass a more expansive perspective (Izaki and Ogawa, 2023).

Although it takes diligent mental commitment, it does not require rigid concentration! Einstein often received creative inspiration while chatting with friends or engaging in mundane activities. Mozart did not shy away from light-hearted partying with friends. Distraction appears to lead to better choices than either conscious thought or an immediate decision (Dijksterhuis et al., 2006).

The Eruption of Creative Insight: Creativity hinges on abstract thinking to distill the crux of a problem, discarding extraneous details. As with all happy moments, creative ideas materialize succinctly in time and present wholesome understanding.

The eruption of insight signifies a mental expansion that heralds the unraveling of a joke's humor, the resolution of a problem, or the formulation of a hypothesis. The mental tension recedes, replaced by a surge of confidence and elation. While refining the finer points might necessitate days or even years, the burst of creativity offers the blueprint for the work's essence, structure, and resolution. Subsequently, we delve into the arts, one of the most renowned conduits of creative expression.

Creativity isn't merely a pursuit; it becomes a fulfilling way of life, enhancing its overall quality.

The Role of the Arts

"Art, freedom and creativity will change society faster than politics."
— Victor Pinchuk

Oscar Wilde considers the ability to appreciate beauty a hallmark of our humanity.[48] Evidence of prehistoric creativity can be seen in sculptures and cave art that date back tens of thousands of years. The earliest Rock Art examples from India, the Middle East, and Germany date back to 700,000 BCE. Although not tangible, music and storytelling undoubtedly were necessary forms of artistic expression. Early shamanic rituals, art, and religion shaped ancient beliefs about the workings of nature, the human condition, and the afterlife. In turn, these core beliefs provided a moral code for human conduct.

The arts provide a kaleidoscope of human experience, as recognized by Aristotle, who understood the power of stories in offering perspective and imagination to improve mental functioning. Perfection[49] is static, cold, and sterile and misses the human element. This is the quality we feel in AI's instantaneous technical ability. Great art conveys positive messages about human potential; its internal rhythm is born from nature, the source of all beauty.

While the audience's interpretation of the artwork may differ from that of the artist (we may never know the prehistoric cave painters' intentions), we connect with the art and the artists through the language of emotions. The unexpected brings forth the self-transcendent and the spiritual, the foreknowledge inspires catharsis, retrospection engenders empathy, and anticipation causes curiosity, tension, and enthusiasm.

The Beauty and the New: Art creates beauty, the symphony of surprising yet familiar. Too much novelty turns the work into bizarre, threatening, or ugly, but lacking surprise is the recipe for boredom. Nevertheless, repeated exposure exhausts art's originality. As Oscar Wilde said, "The moment you think you understand a great work of art, it's dead for you." Therefore, art, the servant of only emotions, not rules, is a mirror of its time.

Surprising elements in art appeal to our sense of novelty, while familiarity provides a sense of security and comfort. The balance between new and safety arises from the emotional balance of temporal gravity (Figure 7). We seek to balance surprise and familiarity in every aspect of life, such as food preferences, housing, and clothing choices. If our creative

---

[48] "Those who find ugly meanings in beautiful things are corrupt, …those who find beautiful meanings in beautiful things are the cultivated."

[49] "There is no great beauty without some strangeness." Francis Bacon.

side is compromised, we become novelty seekers, trying endlessly to catch up to others by shopping for the latest fad or consuming the most sensational news item.

Art's therapeutic and uplifting effects have led to its use in medical clinics, hospitals, and public places.

What is Free Will?

"...free will does not mean one will, but many wills conflicting in one man. Freedom cannot be conceived simply." — Flannery O'Connor

The information value of a stimulus is determined by its potential to surprise. However, our perception and understanding operate involuntarily and automatically; we cannot choose to ignore a stop sign, for instance. What is often termed "free will" is essentially spontaneous behavior. But is free will indeed the central governing force of the brain? Scientific findings suggest otherwise. Extensive research indicates a surge in electrical activity preceding a decision (Libet, 1985). This surge is the readiness potential, signifying the brain's preparation for voluntary actions. Among various choices, the one we make reflects the path with the dominant activation. Free will might manifest as the option requiring the least action, dictated by the brain's energy processes, representing a stationary action.

In the realm of physics, degrees of freedom pertain to the independent parameters of a given state. In the emotional context, this correlates to decision-making freedom. We have free will if we are more powerful than our problems. Going against the prevailing trend[50] restricts focus and autonomy (Izaki and Ogawa, 2023). Within a negative mindset, excessive details cloud clarity, resulting in bias and judgment. Like an electron obeying the forces of its environment, a negative attitude can powerfully control thinking and behavior.

For example, committing a crime or moral transgression is often followed by predictable and highly deterministic behaviors that preclude the notion of free will. The unhappiness that plagues a corrupted mind is akin to substantial momentum, comparable to a ball rolling down a steep slope or a car speeding on a highway. Just as altering the car's course necessitates hitting the brakes, achieving mental freedom requires a mental deceleration. Yet, this deceleration introduces openness and contentment, negating the urge for change (Figure 20). Hence, only a positive mindset truly embodies free will. Paradoxically, possessing free will might diminish the incentive to exercise it.

---

[50]   Doing the opposite of what human nature, expectation tells you to do.

Focus

Determined path — Anxiety

High

Frequencies

Low

Freeedom of associations — Creativity

Joy

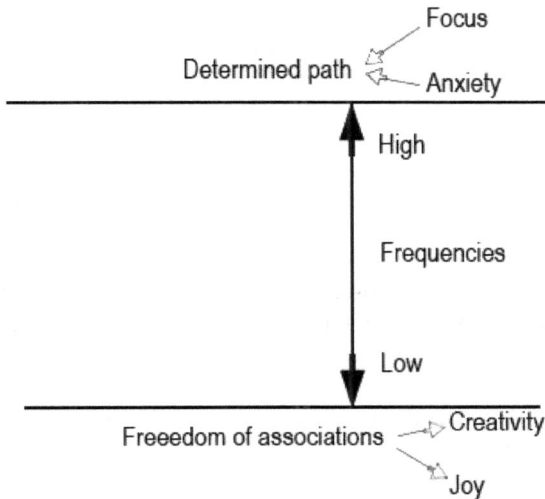

Figure 20. The brain frequencies produce our physiology. Narrow focus is deterministic (no free will). Broad focus permits greater mental freedom, creativity, and joy.

Being practical is essential to effectively engage with the world. There is a notable difference between a person who actively exercises their agency and one who is controlled by compulsions and external influences. If our life is a play, then the world supplies the script. Our attitude consistently serves as a convincing cinematic illusion of control.

Having Free Will but not Being Able to Use It

"We are pushed hither and thither by external causes, tossed about like waves on the sea, and driven by contrary winds, and we never know where we are going." — Spinoza

The phrase "Life, liberty, and the pursuit of happiness" in the United States Declaration of Independence is a testament to the belief in free will. Social governance rests on the assumption of the capacity for voluntary thought and action. While the justice system administers penalties for crimes, criminal intent, or negligence, the defense of insanity, which indicates a lack of awareness of one's actions, can exempt an individual from prosecution. Thus, free will is a profound inquiry with implications for justice, democracy, and equality.

Free will is predicated on the ability to make choices and decisions. However, agency is not just an inherent trait, but a skill that can be honed and enhanced. Like a muscle, it can be strengthened through regular exercise or weakened through neglect. There are instances where it might be stifled, such as in conditions like obsessive compulsive disorder (OCD). If our mental frameworks cater to hidden emotional needs, they reinforce those biases.

Our psychological makeup arises from a 1.5 kg (three-pound) mass of brain tissue. The intricate network of nerve cells generates individual identity and conduct. As observed earlier, automated actions hold sway over our lives, guiding everything from brewing morning coffee to spontaneous smiles. Routine behaviors like brushing our teeth necessitate minimal conscious involvement. Similarly, external stimuli can trigger physiological responses and even emotions without conscious awareness (Uhrig et al., 2016). Because the mind associates these resultant reactions, even those unforeseen or contradictory, with intent,[51] we uphold the belief in the presence of free will. The rationale behind this belief is that the brain is capable of more than its conscious knowledge.

Our range of motion is dictated by the functionality of our bones and muscles, and our accomplishments often align with the social circles we engage with, a phenomenon known as the Matthew effect (Rigney, 2010). While imagination may soar limitlessly, actions carry weight; societal expectations and personal motivations shape our practical autonomy. As Schopenhauer stated, "A man can do as he wills, but not will as he wills."

Internal Organs: Internal bodily processes often evade our conscious awareness, yet our brains consistently receive signals from our internal organs. The rhythm of our heartbeat can influence neural activity, with perception influenced by the heart's cycle. Similarly, the gut microbiome governs cognitive and social functions, mental growth, emotions, and overall well-being (Tooley, 2020).

Parasites and Addictions

The question of free will revolves around the interplay between intention and action. Schizophrenic pathology[52] challenges the grasp of free will, while behavior devoid of internal motivation can resemble that of a zombie. The behavior of a zombie is similar to the implantation of electrodes into animal brains, which empowers scientists to remotely control movement and emotions, analogous to maneuvering a toy car or drone. However, external behavior control evolved millions of years prior via parasitic lifestyle.

---

[51] Initiating, executing, and controlling actions by one's subjective awareness.
[52] Hallucinations, delusions, disorganized thoughts and speech, anhedonia, and lack of motivation are typical symptoms.

Parasites, reliant on hosts for nourishment, safety, and even reproduction, have developed a disconcerting ability for manipulation. Toxoplasma gondii, a single-celled organism reproducing within cats, illustrates this phenomenon. In mice, the infection eradicates the natural fear of cats, facilitating the cat's chance to capture its prey and fulfill the parasite's life cycle. Addictions spotlight an even more distressing absence of conscious presence. While addicts might furnish elaborate justifications for their inexcusable conduct, their actions display a callous disregard for themselves and their loved ones, leading to the loss of homes, relationships, and lives. We will investigate relationships next, which exemplifies the role of the social environment in forming behavior.

The Anatomy of Relationships

"If you can't have empathy and have effective relationships, then no matter how smart you are, you are not going to get very far."
— Daniel Goleman

Only a tiny fraction of the multitude of meteors that collide with Earth's atmosphere yearly redirects back into space. Like these falling meteorites, negative experiences intrude upon our awareness, planting seeds of hopelessness and frustration. Yet, we possess the ability to deflect and deter the undesirable. When our intentions and aspirations outweigh our challenges, we construct a higher trajectory for our ambitions.

Beyond biological needs, such as air and water, emotional intimacy and physical touch are vital for survival. Members of our inner circle significantly affect our behavior and perspectives, involving sentiment contagion. Drawing from a seventy-five-year Harvard study (Ghent, 2011), a direct link emerges between our social connections and emotional and mental well-being. Trusted bonds grant us the courage to reveal our innermost thoughts, fostering vulnerability that positively shapes our quality of life and life expectancy. Close relationships, surpassing social status, IQ, and even genetics, serve as shields against life's woes and contribute to the postponement of cognitive and physical decline.

Although pets cannot replace human relationships, their companionship can ease loneliness. Owners gain psychological and physical well-being from caring for the animal, and pets gain security, support, and interdependence.

Transitioning from Argument to Forgiveness

Both humans and animals can perceive the energy of their counterparts in initial encounters. Vulnerability attracts mistreatment, while confidence commands respect. However, patience, diplomacy, and goodwill should not be misconstrued as deficiencies; underestimating the power of intellect is a perilous oversight.

The Value of Debates: A debate is a discussion in which people support opposing sides of a question. Analogous to a sword fight or boxing match, well-articulated points weaken opponents and invigorate participants with adrenaline. A robust debate enlivens the moment and stimulates thought long after the discourse concludes. Constructive discussions are characterized by respect and cooperation, unlike arguments that aim to assert a point without considering opposing views.

The Nature of Arguments: While debate transpires between parties with respect for each other, an argument implies an emotion-driven dispute. While violence remains an undesirable means of dispute resolution, arguments are equally draining. When conversation partners are entrenched in their own perspectives, they only acknowledge each other when it aligns with their objectives. Analogous to adjusting binoculars' focus, genuine listening requires a conscious effort to comprehend the partner's viewpoint. This type of listening demands mental deceleration. This enriching practice conveys respect, enhancing confidence and goodwill. It kindles the imagination to illuminate opinions, concepts, and intentions.

Addressing Social Anxiety and Facilitating Connection: Social anxiety corrodes self-esteem and trust in interpersonal relationships (He, 2022). Much like static electricity, human interactions generate friction that accumulates over time. This accumulating tension signifies an emotional distance, which can be more agonizing than solitude. Faced with the absence of a familial passionate embrace, individuals instinctively instigate conflict to reestablish the connection. The conflict triggers a spin-up-down pairing.[53] This pairing restores emotional closeness. In longer-term relationships, routine confrontations can evolve into established role-playing.

The Power of Forgiveness: How can we enhance our relationships? (1) Replacing arguments with debates and criticism with forgiveness improves the tone of interactions. Employing more words of praise than discouragement guarantees better connections with everyone. (2) Defusing irritation before it festers into a conflict is essential to preventing

---

[53] According to the Pauli principle, spin-up-down pairing represent minimal energy.

its accumulation. Forgiveness prevents the accumulation of negative attitudes. Adopting a broader perspective fosters compassion that can even retroactively alleviate stress, eliminating the urge for vengeance and retaliation. This practice deepens relationships, fortifies mental resilience, and augments inner strength.

Dignity in Relationships: This respect extends to others when we respect ourselves. Trust extends dignity to humans, emotionally attuned animals, and nature. This trust liberates individuals by bestowing unwavering strength and inspiring admiration. It permits a profound, affectionate, and selfless connection to others.

The Family, from Past to the Future

"Pain travels through families until someone is ready to feel it."
— Stephi Wagner

Certain animal species, like penguins and wolves, mate for life and jointly care for their offspring. However, the human capacity for intellectual prowess gives rise to a family structure that holds intergenerational significance far surpassing that of animals. Families play a pivotal role in upholding culture, behavioral norms, and societal stability.

In hierarchical societies, families acted as strongholds that shielded individuals from natural and social threats such as hooligans, bandits, and local rulers. They served as a form of life insurance for the middle class and impoverished while also functioning as instruments of power and influence for the affluent. Loyalty to the family often precedes individual relevance, and conforming to societal expectations superseded personal viewpoints. Opportunities were frequently constrained by class, caste, and gender, with marriages continually being economic decisions devoid of personal considerations. Those who rebelled risked being ostracized and disgraced. Even within the British Royal family, King Edward's marriage to an 'unsuitable' woman led to his distancing from the monarchy. Curiously, even allegations of sexual abuse have not deterred Prince Andrew's participation in family events.

The Modern Family

From the Stone Age through the Industrial Revolution, scarcities in housing and food, conflicts, and mistreatment were integral to life. Only recently, the family's role, function, and broader social structure have undergone dramatic transformations. Occupations are no longer merely handed down but reflect individual talents and creative potential. The elderly now has access to social safety nets encompassing pensions, healthcare, and end-of-life support, surpassing the care typically provided by family members. The emancipation of women has catalyzed their increased participation in the workforce.

As financial prosperity permits social independence, the size of households gets smaller. In fragmented homes, young parents are often deprived of the experienced assistance that extended family members could provide. Amid their busy lives, parents frequently fail to allocate quality time, even a shared meal, to spend with their children. Perhaps this unprecedented social fracturing is the reason for a surge in addiction, alienation, criminality, and abuse.

Upbringing and Personal Life: Personal history is inextricably woven with upbringing. Beyond physical attributes and family recipes, children internalize their parents' lifestyles and attitudes. Like chromosomes, our upbringing remains an indelible component of our character. Reconciliation with our past, even amid the most challenging childhoods, can propel us toward a positive future. Tara Westover encapsulated this sentiment: "When my mother told me she had not been the mother to me that she wished she'd been, she became that mother for the first time."

Challenges in Family Dynamics

What are the underlying reasons for the failure to raise successful children? One of the gravest forms of parenting failure involves abdicating all responsibility. Allowing children to spend time in solitude, glued to screens, denies them crucial emotional security and affection. The second form of parental failure is creating a tyrannical environment. When children grow up in an atmosphere of fear and mistrust toward their parents, that fear can gradually morph into defiance, aggression, and impulsiveness.

The song "A Spoonful of Sugar" from Mary Poppins harbors a vital parenting truth. Just as we must water the seeds we wish to nurture, we magnify things by granting attention. Praising good and disciplining bad behavior multiplies both equally. Thus, encouragement should precede discipline. Children need consistent, meaningful, positive engagement to fuel their ambition, creativity, and spirit. Families serving as safe nurturing and love havens inspire a more dignified society.

Karma: The earlier discussion on karma focused on the individual. Limitations and micromanagement radicalize children, creating either revolutionaries or weakened personalities. Emotional indifference can produce horrific actions, such as murder and terrorism. Therefore, karma has intergenerational importance, as parents grant the size of their children's dreams. Thus, a person's sphere of influence spreads between a trusted floor (trust in our social environment) and a limiting ceiling (the possibilities provided by our social situation). The secure floor is a trust emanating from peace with our parents, families, and the world. The trust we carry becomes ever more important after losing loved ones. A child with a solid floor can break through any ceiling.

Karma's Wider Influence: In the preceding discussion on karma, the focus rested on the individual. Limitations and excessive control over children's lives can radicalize personalities, spawning either revolutionaries or weakened individuals. Emotional detachment may give rise to abhorrent actions such as murder and terrorism. Hence, karma extends its reach across generations as parents influence the extent and nature of their children's aspirations.

An individual's sphere of impact grows from the solid foundation of trust in their social environment. A secure foundation stems from peace with one's parents, family, and the world. The beliefs we carry assume increasing significance after the loss of loved ones.

The Role of Sleep in Learning

"Sleep is the best meditation." — Dalai Lama

The brain's homeostatic regulation is crucial for learning, attention, motivation, and memory. Sleep supports our physical well-being by facilitating healing and waste clearance. Also, it plays an indispensable role in memory consolidation and learning. Sleep disturbances can have far-reaching consequences, including depression, weight gain, memory issues, premature aging, and even mortality, despite a lack of visible pathological signs.

The Connection Between Homeostasis and Learning: In the sleeping brain, recently activated synaptic connections play a pivotal role. Replaying a diverse range of experiences optimizes the collaborative dynamics of neurons, bolstering memory and learning. Consequently, sleep contributes to both the strengthening and erasure of emotional memories (Yuksel et al., 2023). Emotional memories that remain dormant are susceptible to decay during the rapid eye movement (REM) sleep phase.

However, sleep is not a uniform process; it transpires in distinct cycles. Slow wave sleep (SWS) represents the most restorative phase, characterized by synchronized, slow brainwave activity. Another phase of sleep in REM is when dreaming occurs. During REM sleep, the eyes move beneath closed eyelids, and brain activations mirror wakefulness.

The duration of REM sleep increases throughout the night, constituting approximately 25% of the total sleep duration.

Extraordinary learning capabilities are among the numerous cognitive attributes that set us apart from other members of the animal kingdom. Before birth, generated neural circuits serve as blueprints that anchor experiences after delivery. Babies born with intricate neural mappings have an advantageous head start in life. Sleep further triggers unique patterns of neuronal firing, which can serve as templates for future events. Thus, today's mental state shapes tomorrow's psychological disposition. This connection between current attitudes and future outcomes gives rise to the intricate challenge of aging.

Aging

> *"There are six myths about old age: 1. That it's a disease, a disaster. 2. That we are mindless. 3. That we are sexless. 4. That we are useless. 5. That we are powerless. 6. That we are all alike."* — Maggie Kuhn

Children are born with the ability for joy, anger, sadness, and fear. They observe the world around them, making predictions about future events. Their mental agility enables rapid learning and effective adaptation to significant technological and environmental shifts. Emotional intensity, like boundless joy and profound sorrow, turns the personal lives of teenagers[54] vibrant. Emotional electromagnetism—which can flip in an instant between adoration, despise, passion, and hopelessness—slows the sense of time. Emotional electromagnetism also boosts competitiveness, particularly in boys. Although girls may exhibit less overt competitiveness, emotional gravity imparts a more mature stability. However, it's important to note that individual variations between sexes can be substantial.

Mental Aging: Aging influences our physical bodies, minds, and emotions. With time, individuals rely more on temporal gravity, which tempers the tumultuous emotional fluctuations of youth. Existing attachments and social connections deepen with age, but new ones remain more superficial. While physical discomfort is a source of significant distress for children, the graying of emotions over the years diminishes sensitivity to aches and pains. This emotional attenuation can also hinder learning efficiency and contribute to a sensation of time accelerating.

Our domestic environment shapes our expectations, influencing our social aspirations. People's societal position correlates with their confidence, trust, and aspirations. Educational attainment and work

---

[54] Emotional electromagnetism represents potent emotional polarities that can switch in an instant.

experience act as catalysts for social and professional progression. In middle age, attachments can instigate insecurity, eroding trust, social standing, and influence. This insecurity renders older adults more susceptible to manipulation and deceit.

Dreams are the province of the young, goals characterize the confidence, and memories grace the life of the elderly. As individuals traverse the stages of life, their aspirations evolve, aligning with their age and experience.

In this chapter, we characterized the operational regularities of consciousness, the temporal fermion. The mental particle is identical to other emotional animals, such as birds and mammals. However, our technological achievements put us on a completely different playing field. A forty-hour (and soon thirty-two-hour) work week can provide food, education, housing, and transportation resources, with many hours left for entertainment, social life, and hobbies. Nevertheless, our inability to deal with our emotional demons leads us to addictions that represent temporary modified consciousness or unconsciousness, not real solutions. Regardless, these addictions only make the problem worse over the long term. Instead of benefiting us, our generous access to free time becomes a handicap and a curse.

The next chapter details the consequences of these problems with negative emotions. In contrast, the subsequent chapter offers solutions for mental transformation.

These tools are not impossible or even difficult. In fact, many today already live with purpose and meaning. The lucky few have sharpened their motivation for achievement during their childhood. Nevertheless, this book gives more; it shows that forming responsibility for the common good must be the foundation of society to save our planet. Instead of looking for a home elsewhere, our job is to produce a common future here at home.

For example, a significant outside threat can be the natural trigger for forming a community based on freedom and equality.

Chapter 6

Compromised Performance

"Any person capable of angering you becomes your master."

— Epictetus

The previous chapter talked about emotions' role in normal human physiology, and this chapter focuses on how corrupted emotional regulation leads to personal, relationship, and social problems.

As established in the preceding chapter, the human mind constitutes the smallest unit of intellect. Our innate thirst for learning, creativity, and mental advancement forms the core distinction between us and animals. The intellectual need for progress makes mental stagnation dangerous for our psychology. The mind equates stagnation to regression, which can lead to desperation, addictions, depression, and other psychological compromises.[55] Stress serves as the initial step toward mental degradation.

The middle class can increasingly afford various cosmetic enhancements in the modern era. The obsession with youth inspires diets, drugs, and exercise, but one crucial factor overlooked is emotion. Our emotional well-being can leave lasting marks on our faces, as suffering can be as damaging as heavy sun exposure or alcohol consumption. The following pages delve into the long-term risks associated with negative emotions and toxic relationships.

In the Bible, the story of Adam and Eve presents a narrative starting out in a state of perfect abundance. However, the couple's restlessness led them into rebellion, losing their privileged position. In everyday life, conflict, which often arises from a sense of lack rather than plenty, corrodes people's opportunities even further. Paradoxically, adopting the wrong perspective can lead to frequent clashes with reality, reinforcing the negative mindset that triggered these conflicts in the first place.

---

[55] Diagnostic and Statistical Manual of Mental Disorders (DSM–5), *American Psychiatric Publishing*, 2013.

Corrupted Decision-making

Peter Kropotkin's study of animal behavior revealed cooperation and generosity during times of abundance, contrasted with stress and conflict amid scarcity of resources (Connelly et al., 2017; Kropotkin, 1902). Intellect, generosity, and cooperation require secure and dignified societal structures (Tkadlec et al., 2020), while poverty induces cognitive stress, fostering insecurity and hopelessness. Furthermore, diminished psychological freedom leads to deterministic outcomes, denying free will (Gao et al., 2022).

The thermodynamic understanding of perception shows that managing emotions is like operating a car engine. The intricate integration of emotions within the brain's energy framework means that conflicts trigger physiological arousal (Saunders et al., 2017). Unlike occasional struggle, an inherent aspect of the human experience, chronic stress disrupts the reward and motivation system (Padamsey et al., 2022).

Like using the wrong gear in the car, confused goals, perceptions, and focus cause purposeless activities (Schoeller et al., 2018). Low self-esteem, anxiety, and depression are mental quick-sand that invariably collapse ambitions (Meeusen et al., 2020).

Dopamine is responsible for cognitive flexibility. Suppression of dopaminergic activity blocks the path toward flexibility regions (Hua et al., 2020), which activates the "indirect" stress response instead (Baik, 2020). This response intensifies self-focus (Berkman, 2018) and restricts the degrees of freedom, as depicted in Table I. In turn, a lack of liberty imposes a cognitive burden. In addition, individuals with anxiety exhibit increased connectivity in specific resting-state networks (Gehrt et al., 2022) and regions crucial for social processing, resulting in diminished interpersonal warmth and emotional reactivity (Fredericks et al., 2018).

Impaired emotional regulation correlates with subpar social behavior, IQ (Goldsmith et al., 2020), and depression (Wade-Bohleber et al., 2020). In cases of depression, low entropy insecurity aligns with cognitive vulnerability (Wang, 2020), with disease severity directly proportional to entropy reduction. This low entropy manifests as rigidity (Stringer et al., 2019) and a focus on the past (Zanin et al., 2020).

Overcoming Stress

> *"If the problem can be solved why worry? If the problem cannot be solved worrying will do you no good."* — Shantideva

In the latter part of the twentieth century, improved living conditions afforded the masses more leisure time to engage in social activities, sports, and entertainment. Whether rich or poor, young or old, busyness became a badge of honor, a validation of social significance. However, when schedules are fully packed, unexpected events can swiftly push life out of control.

What is Stress?

Stress is an information flow that invariably accumulates in the resting connection map. We can consider how physical systems respond to energy input to better understand stress. For example, radiation can act as a pump, triggering oscillations, fluorescence, or promoting synchronization (Gentili and Micheau, 2020). Pulsations by flickering light, intricate patterns, and repetitive noises can trigger specific stress responses in people, even if we remain unaware of them. In some people, they can trigger epileptic seizures (Fisher et al., 2022),

Similar to mechanical stimuli, repetitive information can provoke anxiety. Although our subjective tolerance to stress varies, information overload triggers the sense that time is slipping away, causing impatience and impulsiveness. Thus, stress functions like a pump, fueling emotional swings, fluorescence, or synchronization. For example, passive aggression, analogous to fluorescence, occurs in close relationships. Additionally, stress can synchronize behavior within a group, contributing to mob behavior or social lasing. These varying effects of stress will be explored in detail in subsequent sections. Therefore, stress is an information radiation, and its predictable results are within the realm of fermionic response, underlying the mind's particle nature.

Varied Effects

Stress mobilizes protective mechanisms against danger. Adrenaline and cortisol boost our ability to respond to challenges by increasing heart rate, respiration, and blood pressure. Nevertheless, these defensive hormonal changes can adversely affect the immune system, memory, and cognitive processes. Sleep disturbances, weight fluctuations, and digestion issues can undermine motivation and hinder performance. Stress wide-ranging health consequences make it a significant concern in both professional and public health contexts.

Analogous to a small window frame, stress narrows our ability to perceive a broader perspective. Nevertheless, the same stressor can impact individuals differently. Its subjective nature underscores its connection to self-confidence and social support. Although stress transcends social boundaries, it is perilous in competitive environments like sports and high-pressure jobs. Shame serves as a poignant example.

The Power of Shame

Social support is crucial in nurturing self-esteem, providing a foundational sense of security that fosters accountability. In such a secure position, withdrawal of social validation can lead to intense feelings of shame. Higher social rank potentiates shame's falling or shrinking sensation, potentially propelling self-directed or outward aggression (Vries, 2017). Self-loathing triggers self-deprecation, compulsive behaviors, and even addictions to escape these emotions. Fixating on others' opinions can breed feelings of inferiority, envy, jealousy, falsehoods, and criminal behavior.

Outsiders, including the homeless, tourists, and immigrants, lack social support. For them, it is easy to disregard social norms, which renders them somewhat immune to shame's effects.

Managing Chronic Stress: We are the first generation to analyze scientifically how our feelings influence our well-being. Rather than eradicating stress, we must learn to manage it by confronting challenges rather than being intimidated. Each successful response expands our mental horizons, revealing opportunities for growth and development. The subsequent section delves into negative emotions and their management.

Negative Emotions

"Again and again I have said to myself, on lying down at night, after a day embittered by some vexatious matter, 'I will not think of it anymore! . . . It can do no good whatever to go through it again. I will think of something else!' And in another ten minutes I have found myself, once more, in the very thick of the miserable business, and torturing myself, to no purpose, with all the old troubles." — Lewis Carroll (1893)

Throughout the annals of human history, a grim narrative prevailed. Staggering levels of violence and subjugation characterized labor and life conditions. Even in eras marked by peace, the yoke of power inequalities has weighed heavily on society. Now, at a time when technological and social advancements offer unprecedented opportunities for human growth, we find that the emotional legacies of our fraught history continue to compromise our collective aspirations.

When confronted with physical threats, our bodily protective instincts seek security in separation. This opposition is analogous to the opposite spin configuration of the Pauli exclusion principle in quantum mechanics (Figure 10). Muscles tense to protect vital organs, and the "fight or flight" response activates, ready to flee from danger. While this evolutionary adaptation ensures immediate survival, it is detrimental in the long run. This primitive, self-centered focus narrows our mental bandwidth, diminishing our free will and decision-making capacities. The insecurity fuels impulsive and contradictory behaviors, echoing the quixotic endeavor of moving the world without a fulcrum. As research by Nuno-Perez et al. (2021) indicates, the residual effects of these negative emotions distort our intuition and judgment, pushing us into a cycle of pessimism.

The Pitfalls of Insecurity: Stress, consequently, causes insecurity, acting as a damper that compromises performance or entirely prevents action. In the first case, it prompts individuals to undertake challenging projects as a means of validation. However, insufficient planning and effort management usually culminates in depleted focus, especially during critical junctures like essential examinations or vital presentations, exacerbating self-doubt.

On the other end of the spectrum are those so paralyzed by insecurity that they can't even commence a project, resorting instead to procrastination and excuses. The psychological burden of these conflicts further isolates people, depriving them of meaningful emotional connections and reinforcing their pessimistic outlook. When the mind's temporal compass is thus skewed, it becomes directionless, unable to orient toward progress.

Health Consequences

> *"You are already a star; you have nothing to prove—the need to project confidence burns out your core. Over time, pride, anger, and self-assertion exhaust your energies, and you collapse into your shadow-self."* — Eva Deli

Pride, anger, and self-assertion are draining for one's emotional reserves, degrading resilience. While negative emotions are crucial in self-preservation, their prolonged presence drains our energy and undermines free will. Aggression accelerates cognitive decline and leads to apathy and other adverse health complications. Negative emotions are for the benefit of the moment, but dwelling on them is detrimental.

Newton's First Law of Motion explains the psychological inertia carried by human habits: a body at rest tends to stay at rest, and a body in motion tends to stay in motion. Habits, thus, can propel us toward a brighter future or ensnare us in a cycle of detrimental behaviors like addiction and inactivity. Ironically, these damaging habits often originate from the issues—such as low self-esteem—that they exacerbate. Because emotions arise involuntarily, altering these negatively ingrained patterns demands a nuanced understanding, almost like decoding a complex geometric theorem like the Pythagorean principle.

The Versatility of the Pythagorean Theorem: From Geometry to Human Psychology

"Be willing to have it so. Acceptance of what has happened is the first step to overcoming the consequences of any misfortune."
— William James

The Pythagorean theorem[56] is a cornerstone concept in mathematics (Figure 21). Originating independently in cultures as diverse as Mesopotamia, China, and India, it has been a staple of human intellectual heritage for millennia. This theorem posits a relationship between the squares of the sides of a right triangle and its hypotenuse. Intriguingly, when the hypotenuse remains fixed, the lengths of the other two sides display an extraordinary interplay: as one approaches zero, the other surges toward infinity.

The generality of orthogonal relationships gives Pythagoras' theorem broad utility. It can predict architectural measurements, engineering stress analysis, and quantum uncertainties. With appropriate considerations, we can extend its relevance to psychology.

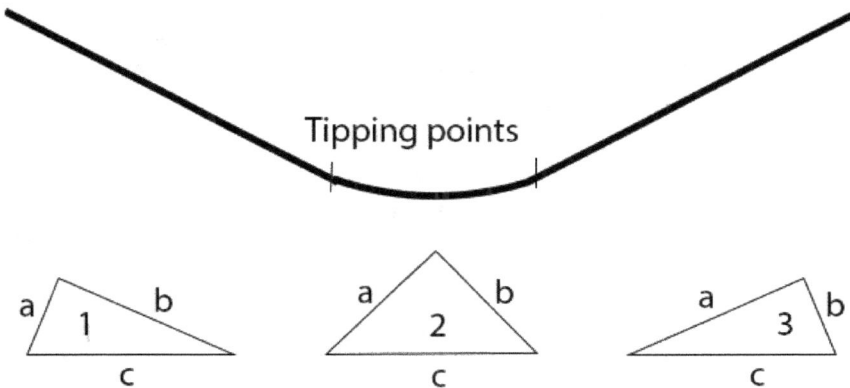

Figure 21. Orthogonal relationship of the right triangle. The three right triangles (1,2,3) have the same hypotenuses. The two sides (a, b) of triangle 2 are similar in length. As the sides' discrepancy increases, the differences in their squares grow exponentially (1 and 3).

---

[56] The Pythagorean Theorem states that the square of the hypotenuse is equal to the summed squares of the two sides.

Pythagoras and the Mind

Earlier, we discussed the orthogonality between the spatial sensory input and the temporal nature of cognition. We also applied this perpendicular quality between wave functions and fields to the psychology of attention. Motivational forces, represented as vectors, can be directed toward the past or the future. Past motivation is a futile exercise that CANNOT bring results. Consequently, attention is our most valuable asset, best directed toward what uplifts and fulfills us.

Even in social contexts, past focus is detrimental to companies and nations. It presents solutions with limited degrees of freedom, excluding forward vision. Russia's dream of restoring lost territories is supposed to hide the regime's increasingly dictatorial leadership style.

Physical challenges that exceed our capacity risk injury, while too simple tasks lead to boredom. The sweet spot lies in aligning tasks with our abilities, thus generating a state of flow tailored to our unique skills and mental readiness.

The Power of Acceptance

Personal growth thrives in an encouraging atmosphere, whereas a stifling environment smothers ambition. However, changing the focus of our attention can powerfully influence our beliefs; thus, acceptance can potentially convert negative experiences into opportunities for growth.

As governed by the Pauli principle, personal space depends on one's mindset. Anxious people zealously guard their space, reacting strongly to perceived encroachments and attempting to create substantial emotional distance from others.

The Transformative Power of Surrender: While self-defense is a natural human reaction, insurmountable difficulties often lead to capitulation (Figure 22). Surrender is the realization of the hopelessness of struggle. When we relax our guards, the mind accepts the situation rather than fights it. Because the thermodynamics cycle of cognition is reversible, yielding to the stronger forces reverses the cycle direction. Therefore, surrender represents fresh perspectives and deeper understanding originating in organic solutions.

Success comes from converting hardships into experiences that contribute to mental evolution. Acceptance, by tearing down barriers, fosters societal unity and advances communal wisdom. We delve deeper into the nuances of social support in the final chapter, including actionable recommendations for implementation.

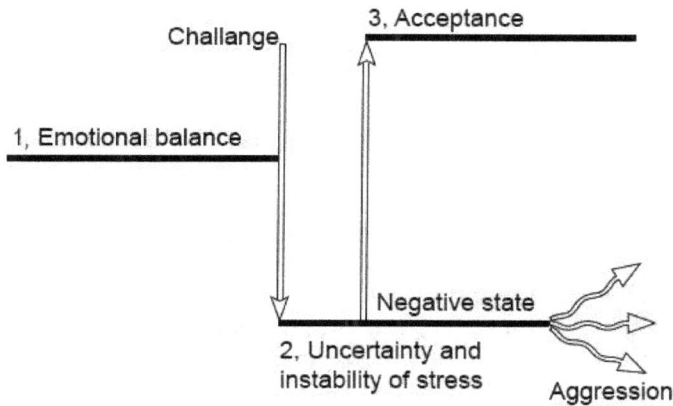

Figure 22. Possible outcomes of stress and acceptance. Black horizontal lines represent the emotions' energy level. Disturbance unbalances the mind (1) and causes stress (2) Cynical, bitter memories accumulate into an avalanche of anger, criticism, or violence. Erasing the stress via acceptance (3) restores flexibility.

The Multifaceted Role of Pain: From Biological Signals to Emotional Scars

"Pain changes people, it makes them trust less, overthink more, and shut people out." — Unknown

Pain, a typical symptom in many medical conditions, has a varied and vital function. At its simplest, it acts as an alarm system indicating potential harm and prompting us to take evasive action. While acute pain can be immobilizing, milder forms often spur us into exploratory behavior as we search for alternative ways of movement that avoid aggravating an injury.

Treatment for Chronic Pain: More Than Just Relief

Pain is an effective survival tool but depletes emotional and physical energy when it becomes chronic. One of the most draining aspects of enduring pain is the psychological weight of its seeming permanence. Hence, the reassuring words "the pain will soon go away", which lifts the burden of infinity, can bring immediate emotional relief. The belief that emotional suffering is also transitory is an equally liberating thought.

Therapeutic Approaches: Pain management drugs, while effective, often come with significant side effects and risks of dependency. Hospitals increasingly turn to alternative therapies—yoga, psychological interventions, hypnosis, and cognitive-behavioral therapy—for relief.

Methods like music therapy, art activities, or simply looking at photos of loved ones can enhance pain tolerance. Awareness of pain's physical qualities can also diminish its subjective intensity.

The Ripple Effects of Suffering: Suffering extends beyond immediate pain, incorporating the memory, anticipation, and dread of it. Therefore, suffering multiplies the pain, resulting in a numbing effect, which can spiral into apathy and even lead to conditions like depression and addiction. Sharing our struggles can offer therapeutic value. Experiencing empathy, whether by understanding others' pain or through cultural means like literature and film, enriches our emotional repertoire; "feeling the pain of others" helps us to become better people.

The promise of treatment is already a powerful physiological remedy.

Existential Pain and Emotional Scars: Pain also includes emotional and psychological trauma, which can be as indelibly imprinted in our minds as a photograph. For example, post-traumatic stress disorder (PTSD) can trap individuals in a temporal loop of reliving traumatic events. Like arresting movement to avoid bodily pain, the brain resorts to emotional avoidance to escape psychological trauma. However, attempts to suppress such thoughts often have a counterproductive effect; Lenz's law explains how thought suppression leaves the emotional pain ever more in focus.

The biblical Eve, Orpheus and Eurydice,[57] and many other literary examples depict the destructive outcomes of repressed emotions. Studies indicate that emotional suppression amplifies pain and anxiety and can trigger depression (Wenzlaff and Wegner, 2000; Ghandehari et al., 2020). Crafting an honest and meaningful life story can negate the need for thought suppression, leading to a healthier emotional state.

Understanding pain in its multifaceted forms—physical, emotional, and existential—enables us to approach it holistically. This understanding can inform treatment options and provide insight into our shared human experience.

The Role of Belief in Pharmacological Response

Our understanding of the brain has traditionally been rooted in neurochemistry and structural biology, emphasizing the roles of neurotransmitters, synapses, and neuronal pathways in modulating behavior, cognition, and emotion. Pharmacological drugs lead the way

---

[57] In trying to free his recently deceased wife, Orpheus visits the Hades, the ruler of the underworld. He reaches an agreement; he can free her if he walks back to the surface without looking back. The task is impossible for Orpheus; he cannot suppress his burning desire to see his beloved wife, so he loses her forever.

to understanding how external substances can alter brain function. Yet, emerging research suggests that nonpharmacological constructs, particularly beliefs, might also profoundly influence brain activity, perhaps even in a dose-dependent manner akin to drugs.

Initial evidence comes from addiction science. Studies have shown that drug abusers display differential brain activity based on their anticipation of drug enjoyment. For instance, if they expect a drug, their perception of its pleasure is heightened (Kirk et al., 1998; Volkow et al., 2003). The psychological construct of 'expectation' significantly modulates their neural responses, potentially priming their brains for a more potent reaction to the drug. Similarly, in pain management, the very belief in a medication can shape its efficacy. Views about pain medications can cause pain catastrophizing, exaggerating pain experiences (Elphinston et al., 2021).

Moreover, beliefs also seem to play a crucial role in our immune responses. A startling revelation is that non-pharmacological variables, specifically beliefs about vaccines, predicted nearly a third of vaccine side effect severity (Mattarozzi et al., 2023). Emerging research indicates that beliefs influence brain activity in the same systematic, dose-dependent manner as drugs (Perl et al., 2023).

For instance, nicotine increases activity in the thalamus (a nicotine-binding site) depending on belief intensity (Perl et al., 2023), similar to the dose-response curves seen with many drugs. Furthermore, robust beliefs about nicotine increase the functional connection between the thalamus and the prefrontal cortex. This region evaluates rewards and risks.

Such findings hint at a revolutionary perspective: Beliefs can impact neural circuits in ways previously reserved for drugs. This challenges the traditional separation between 'chemical' and 'psychological' influences on the brain, suggesting a more intertwined relationship.

Therefore, beliefs, long thought to be abstract constructs, have tangible, measurable effects on the brain, potentially reshaping our understanding of addiction, pain management, and even immunology. This underscores the importance of a holistic approach to medicine and recognizing the influence of placebo.

The Power of the Placebo

"I would rather know the person who has the disease than the disease the person has." — Hippocrates

Modern Medical Advances and Their Limits: The advent of surgeries, pharmaceuticals, and vaccines has revolutionized our approach to illness, offering effective treatments for many conditions. However, the mental aftermath of disease—symptoms like depression, anxiety, and fatigue—

poses a more complex challenge for healthcare professionals. Clinicians often observe a distinction between patients; those who actively engage in their recovery process often get well, and those who remain ignorant or fake confidence often succumb to their conditions.

The Mind-Body Connection Unpacks the Psychological Dimensions of Illness and Healing: The placebo effect exemplifies how individual differences contribute to healing without directly altering the disease. Genetics, psychological states, brain chemistry, attitudes, and social context influence recovery. The environment in which medication is administered, societal expectations, and a patient's belief in the efficacy of the treatment also play significant roles. Transparency and genuineness, both in interpersonal interactions and medical settings, positively affect the experience of illness. Approaching their disease and treatment options with an open mindset reduces patients' pain levels and boosts placebo impact.

The Nocebo Effect: The Dark Side of Psychological Influence: The "nocebo effect" undermines the effectiveness of medical treatments through adverse psychological mechanisms. Conditions like depression, disrupted sleep, poor appetite, and significant blood pressure drops can result from this effect. Anxiety about undergoing treatment can manifest as physical symptoms, including pain, nausea, and gastrointestinal problems. Patients who fixate on the potential side effects of medication may inadvertently bring about those very symptoms, making their fears a self-fulfilling prophecy.

The Intersection of Mind and Body: The placebo and nocebo effects underscore the extraordinary interplay between our psychological state and physical well-being. These effects reveal that our bodies are not just passive recipients of medical treatments; our beliefs, attitudes, and emotional states can significantly influence how we respond to illness and healing.

Understanding the psychological dimensions of medical treatment provides a more holistic view of patient care. It highlights the need for an integrated approach that considers the physiological aspects of illness and the critical role that mental and emotional states play in recovery.

Emotional Temperature and Pressure: The Phases of Matter as a Lens for Understanding Social Dynamics

"Anger comes from pain, and it causes pain." — Le Guin

Transitions of Social Phases: Solids, liquids, and gases are the familiar phases of matter in our everyday life. Phase transitions can transform solids into liquids and liquids into gases in a process that can also take place in reverse order. Social formations can also undergo similar phase changes. These shifts can be gradual or abrupt, much like how physical properties can suddenly alter during a phase transition. For example, the sudden societal shift closely aligns with the concept of mob behavior, which we will explore further in the next section.

Social Solids: Stability Through Strong Attachments: Strong, stable relationships form the building blocks of "social solids", characterized by emotional depth and durable bonds that can be insulators or conductors. However, increasing social or emotional 'temperature' can introduce more fluidity into these solid structures.

Democratic Societies Represent Social Fluids: In the liquid phase, particles occupy a defined volume but move freely, offering greater flexibility. Similarly, democratic societies behave like fluids, where flexible interactions enable the diffusion of attitudes across communities.[58] According to recent studies, fluid dynamics models can accurately describe opinion dynamics (Yang et al., 2020). In this analogy, the capacity to influence people corresponds to the liquid's height. At the same time, collective mood or sentiment equates to temperature.

The Universal Gas Law

At a sufficiently high temperature, heated liquids can transform into gases, allowing particles to bounce freely in three dimensions, endowing them with high degrees of freedom. Likewise, in a social context, mistreatment can reach a critical point when people assume volatile and confrontational behavior (Figure 23). The Universal Gas Law,[59] which describes the behavior of gases in terms of volume, pressure, and temperature, can provide insights into this societal dynamics. Although emigration can ease social pressure, societies with well-defined social temperatures have limited expansion possibilities. In such fixed societal 'volumes', emotional or psychological tensions boost social 'temperature', causing conflict or unrest. What can the universal gas law tell us about human psychology?

Emotional Equilibrium and Homeostasis: This pressure is often a stable trait that can loosely indicate one's social class. For example, growing up in stressful settings can lead to impulsive behavior or paranoia. When conflicts are the exclusive notable events in a child's life, peaceful settings feel uninspiring. Engaging in conflict becomes a self-reinforcing cycle that contributes to an individual's sense of excitement or purpose.

---

[58] Totalitarian governments create an atmosphere of fear, effectively freezing the social behavior.

[59] Universal gas law describes the relationship between an ideal gas' volume and pressure at a specific temperature.

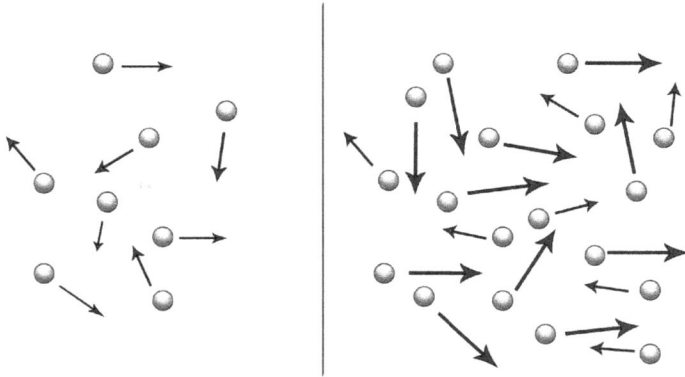

Figure 23. The relationship between particle collisions and pressure. Molecules form low pressure and slow particle velocity (left). Increasing the number of particles (or their speed) creates compression, increasing temperature (right).

Subjective Well-being Homeostasis[60] shows that we form a stable psychology around a set homeostatic equilibrium (Diener et al., 1999). As a result, we may start activities with no other use than to generate conflict and tension. Confrontation triggers stress and resolution, which satisfies the need for a stimulating environment. Understanding our social comfort zones is critical to navigating the complexities of human behavior and relationships.

·   *Consequences of the Gas Law for Society*: Just as varying pressures and temperatures can produce unique states in physical systems, fluctuations in emotional 'temperature' and 'pressure' can lead to nuanced and complex social formations. Transitions between these states can be dramatic and emotionally taxing, akin to the energy consequences of phase transitions in physical systems. These changes can raise fundamental questions about individual self-worth, creating emotional states that are as puzzling and unpredictable as non-Newtonian fluids.[61] In analogy, sometimes, people are unable to produce intelligible responses immediately. Such freezing behavior is analogous to a non-Newtonian fluid.

Our emotions reveal with clarity our range of social comfort.

The Psychology of Mobs and Social Uprisings

Boiling is the transition of a substance from liquid to gas, increasing the particles' kinetic energy and range of motion. The boiling point is the

---

[60] Subjective Well-being Homeostasis shows that each person has an automatically maintained life satisfaction point.
[61] Non-Newtonian fluids change their behavior under stress. For example, a mixture of corn flour and water responds to a force by developing very high viscosity momentarily.

temperature during boiling, which can change with surrounding pressure. The endothermic process can provide a surprisingly valid analogy for social science protests, rebellions, and revolutions.

Economic pressure and social tension increase the social temperature for masses of people. Because compression increases the boiling point in a liquid, oppressive regimes use overwhelming force to suppress public displays of discontent and protests. However, releasing the pressure triggers a sudden and violent explosion into gas. Indeed, historically, uprisings often happened during the easing of oppression, such as in France in 1789, Russia in 1917, and China in 1949.

Behavioral Dynamics in Uprisings: During social unrest, individual behaviors often amalgamate into a volatile, collective action. French social psychologist Gustave Le Bon observed during the 1848 Paris Rebellion that the behavior of crowds differed vastly from that of individuals who comprised them. Once the social temperature reaches a "boiling point", collective actions can become dangerous and unpredictable, free from individual restraint or control. When the energy dissipates, the crowd disbands, leaving individuals to grapple with the aftermath, either taking pride in their collective actions or feeling shame for losing control.

Behavioral Hysteresis

> *"The line between good and evil is permeable and almost anyone can be induced to cross it when pressured by situational forces."*
> — Philip Zimbardo

Hysteresis is a memory-dependent behavior in elastic materials and ferromagnets. This concept is well-studied in materials science and applies to psychological states. For example, the particles line up in stretched-out flexible materials, warming and compromising elasticity. The material must cool down before its elasticity recovers (Figure 24).

In psychology, hysteresis explains how the perception of a stimulus is influenced by immediately preceding perceptions. The perception is gradually modified during successive changes until it is perceived as a different category. However, the decreasing stimulus values produce another crossover point when the changes occur in reverse order. This phenomenon explains why people are slow to get angry but even slower to calm down.

Hormonal Regulation and Emotional States

Hysteresis in Society: In terms of emotional responses, small fluctuations in hormonal levels like testosterone and cortisol can lead to disproportionate

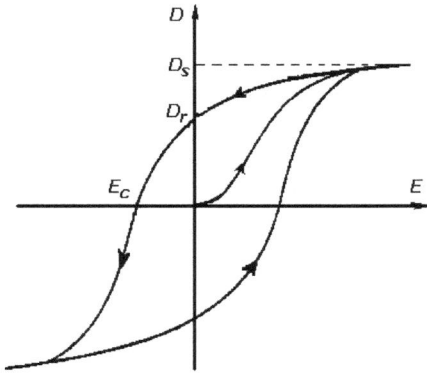

Figure 24. Hysteresis The accumulating energy forms a memory, which changes the response. The system's altered memory in the reverse process creates a different trajectory.

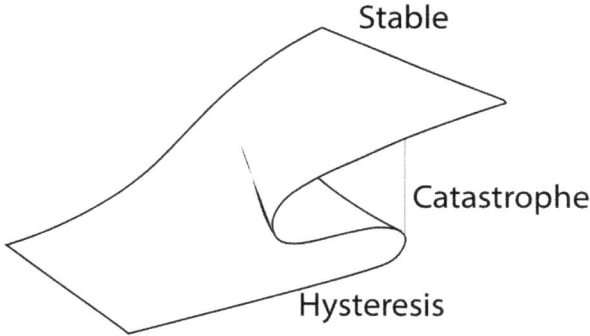

Figure 25. Emotional hysteresis of a dog. The behavioral surface has a single solution on the far left and the far right. The folded region represents memory-dependent performance. The dog can switch between angry and fearful (see fight and flight response in Figure 10). The lower situated fold (open to the left) represents flight triggered by fear. The top fold (on the right) represents a fight powered by dominance in motivation. A catastrophe is a sudden jump in behavior. Within the folded surface, even a tiny shift produces substantially different outcomes.

changes in states of aggression or fear, dramatically disrupting the smooth transition between the relaxed and angry state (Figure 25). Testosterone generally increases aggressive behavior ('fight'). In contrast, cortisol impels fear ('flight'), even at high testosterone levels (Knight et al., 2019; Mehta and Josephs, 2010). Once an emotional state is triggered, it takes significant time for an individual to return to a baseline, calm state.

Economic Hysteresis: Hysteresis also applies to economic scenarios. In economies, the trajectory differs when unemployment increases or decreases during the economic recovery. Hysteresis explains that unemployment rates tend to linger even after a recession has technically ended. Various factors contribute to this inertia, including investor pessimism and a workforce that may have seen its skills degrade or become obsolete during prolonged periods of unemployment.

The Mob as a Form of Hysteresis: In a societal context, mob behavior can also be viewed through the principle of hysteresis. Once conflict escalates to a certain point in a crowd, bringing it back to equilibrium can be challenging. This lag, or hysteresis effect, makes mob behavior unpredictable and sometimes dangerous.

Social Ferromagnetism

The substantial strength of electromagnetism compared to gravity comes into play in bipolar materials.[62] When arranged in a lattice, each spin interacts with its neighbors[63] (Figure 26), forming an ordered, magnetized, and stable state. Ferromagnetism is bipolar materials' ability to align with an outside magnetic field into a quasi-bar magnet. Although some ferromagnets can maintain the magnetism indefinitely, arbitrary particle movements usually restore the disordered atomic orientation.

Emotional electromagnetism represents strong bipolar motivation, causing our attitudes to alternate between attraction and despise. As a result, it can trigger "Social Ferromagnetism," an alignment of individual beliefs and behaviors within social and cultural 'fields'. The theory suggests that people are highly influenced by their social circles, adjusting their opinions and actions to align with those of the group. Our social antennas turn us into ferromagnets, absorbing our social circles' intuitions. Subtle bodily signals and intonation (prosody) can carry emotional messages without the speaker's intention or awareness. Emotional nuances, often unintentional, are communicated between individuals and can affect a group's collective mood or perspective.

A shared orientation reflects the social-political climate and the members' perspectives. People in agreement intuitively copy each other's body position; social attunement can lead to sentiment contagion, a subconscious transfer of political views, happiness, or discontent. This theoretical framework can explain riots or mob actions as phase transitions in magnetic materials where increased thermal energy disrupts the ordered magnetic alignment, leading to a disordered state.

The "Free Will" section also explores the correlation between individual attitudes and social belonging.

---

[62] In bipolar materials, the charges are not symmetrically distributed, causing them to act like a bar magnet.

[63] The Ising model can represent the dynamics of magnetization.

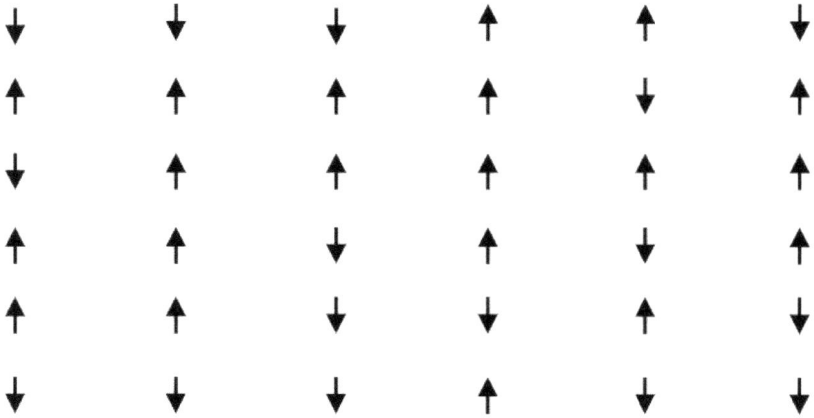

Figure 26. The Ising model. The Ising model represents magnetic dipole moments (spins) in one of two states (+1 or −1). The spins are arranged in a lattice (where the local structure repeats periodically in all directions), allowing each spin to interact with its neighbors.

Filter Bubble

The "Filter Bubble" concept focuses on how social media and search engines can further entrench people's preexisting beliefs and attitudes. This algorithmic personalization restricts exposure to differing viewpoints and can enhance collective bias within a group. The information individuals receive aligns with their preexisting beliefs, thereby radicalizing the members' collective bias and creating filter bubbles.[64]

Emotion-driven information can trigger transitory swings in opinions, inspire protests, shift election results, or inspire social change (Khrennikov, 2020), as shown in social lasing. The following section discusses the emotional ping pong, called "Emotional Fluorescence", also known as passive aggression.

Emotional Fluorescence

> *"Don't hold on to anger, hurt or pain. They steal your energy and keep you from love."* — Leo Buscaglia

We often consider the atomic structure a mini solar system due to the electron's quantized angular momentum (Figure 27). Although the electron cloud is highly dynamic, individual particles can only occupy a discrete energy orbital, which makes them vulnerable to incoming photons.

---

[64] Filter bubble is a state of intellectual isolation of internet users, which excludes contradictory information.

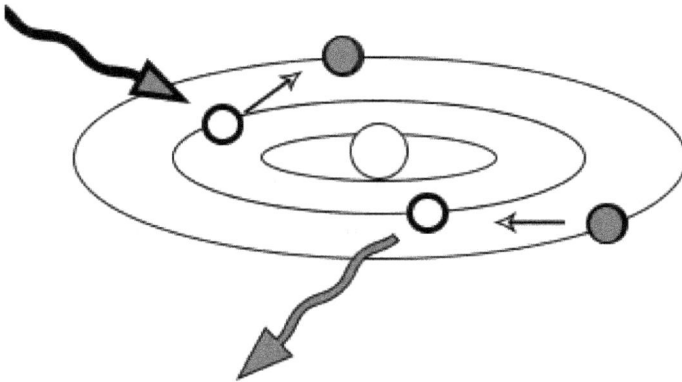

Figure 27. Fluorescence in an atom. The concentric ellipses indicate the electrons' energy level; the central circle is the atomic nucleus, and small balls represent the electrons. Collision with a photon (gray incoming arrow) pushes the particle into a higher, unstable energy orbit (gray). When the absorbed energy radiates out, the electron falls back into its original energy state. Visible radiation is called fluorescence.

High-energy photons can kick the electron out of the atom. Still, typically, they push them into an elevated, unstable energy orbital. Eventually, the particle returns to its original position and emits the borrowed energy. In exceptional cases, the process is a spectacular visible radiation called fluorescence.[65]

Like the atomic world, society is highly structured, where individuals form stable attachments to relatives, friends, and even organizations. In turn, these attachments represent distinct energy levels, which give rise to our social hierarchies akin to the discrete energy levels of an atom. Nevertheless, people are continuously exposed to social energy. When information radiation affects people with strong social attachments, it boosts the resting state energy level (Xu et al., 2023), pushing it into a more energetic disposition. Emotional distance or detachment dictates behaviors like the "silent treatment," also called passive aggression (Gao et al., 2022).

The absorbed energy reduces mental freedom. Like a caged dog that walks up and down to find the exit, the stressed mind produces thoughts that move back and forth to find a different outcome. However, the past has already happened; thus, a solution cannot be found. Eventually, the tension of the borrowed energy snaps, resolving itself through deterministic and impulsive behavior (Aberg et al., 2023; Lin et al., 2022).

How does passive aggression originate from the mind particle organization? Newton's third law, which states that every action has an

---

[65] Fluorescence is the visible light emitted by electrons after being subjected to radiation of a shorter wavelength light.

equal and opposite reaction, is analogous to how stress operates within this context. Unhappiness triggers memories that validate their presence and foster a desire for retribution, as illustrated in Figure 27. Emotional fluorescence[66] is a theatric retaliation executed under the banner of perceived 'justice'. This release of built-up tension frees individuals from more severe psychological anguish and depression. The outbursts, which can target relatives, friends, or even innocent bystanders, are similar to avalanches in a sand pile.[67] Therefore, the timing and intensity of these emotional releases remain unpredictable, contributing to the increasing mistrust and division.

Passive aggression, followed by subsequent reconciliation, can coexist within stable partnerships. Analogous to a pendulum's movement, negative emotional energy can oscillate regularly between couples and friends, detrimentally impacting interpersonal trust. If such attacks are one-sided, they fall under the categories of bullying or abuse, explored in the following sections.

Harassment and Bullying

> *"People who love themselves, don't hurt other people. The more we hate ourselves, the more we want others to suffer."* — Dan Pearce

Regardless of societal standing, relationships founded on mutual respect foster collaboration and generosity, while those built on power disparities lead to competition and mistreatment. When intimidation happens outside of a personal relationship is commonly termed bullying. This coercive need to control others can manifest through cyberbullying in various settings, including schools, workplaces, or online. Discrimination based on race, religion, gender, age, disability, or nationality constitutes harassment.

Harassment and bullying are the tools of insecure people. These acts serve as a quick, albeit temporary, boost to their self-esteem. Bullying often stems from preexisting advantages, including physical strength, social standing, financial status, race, gender, or religious beliefs. This skewed power dynamic results in a preordained outcome that elevates the bully at the expense of the vulnerable target. In severe cases, sustained bullying can lead to learned helplessness, as the constant sense of powerlessness erodes one's social motivation.

While demeaning others may earn short-term approval from like-minded individuals, bullying's fleeting and shallow victories ultimately undermine meaningful social relationships. The cycle of

---

[66] Emotional or mental fluorescence is criticism, aggravation or physical violence that releases accumulated emotional energy.

[67] In the sand pile model, the addition of each sand grain can trigger an avalanche.

mistreatment often cascades downward through society, becoming concentrated in underprivileged communities. Bullying can perpetuate in a vicious cycle because bullies often were abuse victims during their formative years. However, it's worth noting that bullies can sometimes overreach, becoming the recipients of the same abusive behavior they once meted out.

The most effective way to combat bullying and harassment is to foster a community culture grounded in trust, respect, and accountability. Bullies typically avoid those who exude confidence and emotional stability, validating the notion that prevention is more effective than cure. Echoing Eleanor Roosevelt, "No one can make you feel inferior without your consent." We play an active role in shaping how others treat us, and the signals we send inevitably influence the reactions we receive from our social circles.

Abuse

"My father was 'a merchant of chaos', a ''bully' and a 'coward' who beat his children." [My father] was the kind of person where, if something goes wrong, they kick you. It was a great lesson in my life—how he'd lull you in, make you feel safe and then, bang!" — Tom Cruise

To the casual observer, bear dancing may seem quaint and harmless. However, it conceals a methodic cruelty, starting with the abducting of a bear cub from his mother. Trainers subject the cub to extreme pain by placing it on scorching hot metal sheets while playing music. The desperate cub tries to escape the deadly burns by lifting his paws one after another, forming a primal association between music and dancing. The indelible link in the animal's mind between the music and the dance reflex is a conditioning that destroys the bear's ability to live freely.

Similarly, different forms of exploitation—verbal, physical, sexual, and psychological—leave enduring emotional scars on humans. Practices like slavery, legally sanctioned as recently as 150 years ago, profoundly undermined individual autonomy and the drive for independence. Even after the formal end of such practices, the residual effects of segregation and colonization linger. Tragically, the darkest facets of human nature often manifest in the abuse of family members, including children, spouses, and the elderly.

Like bullies, abusers are driven by a need for dominance and control. Still, their behavior often originates from underlying fears and insecurities. The abuser cultivates emotional closeness with the victim but inflicts pain timed to a sensitive or vulnerable moment. This way, the abuse shatters trust and inflicts emotional wounds, systematically eroding the victim's spirit.

Abuse engenders a toxic dependency, akin to the dancing bear's conditioned behavior, progressively destroying the capacity for escape. Adult survivors of abuse often find themselves navigating relationships that mirror the ones they knew as children, perpetuating a multi-generational cycle of mistreatment. Overcoming the lingering emotional impact of such abuse can liberate future generations from its curse.

Physical Abuse: Any form of abuse is a short-term tactic with long-lasting repercussions, notably stunting emotional and intellectual growth (Figure 28). The betrayal of receiving pain from those expected to offer support warps human psychology and hinders healthy emotional development. A 1993 UN report labeled domestic violence as a public health crisis, highlighting that corporal punishment like spanking[68] can foster aggression. In response, many countries have enacted legal protections against family violence, including banning corporal punishment in schools and homes.

Emotional Abuse: Unlike physical abuse, emotional abuse erodes trust and damages the psyche without leaving visible physical signs. The inability to establish healthy boundaries makes individuals susceptible to maltreatment. Children, wholly dependent on their caregivers, are particularly vulnerable to emotional abuse, including belittlement, suspicion, and rage. The more harm an abuser inflicts, the greater the control he gains over the relationship. Such emotional trauma can result in self-esteem issues, leading to emotionally withdrawn or volatile personalities (Figure 29). Abuse hardens people, forcing them not to trust

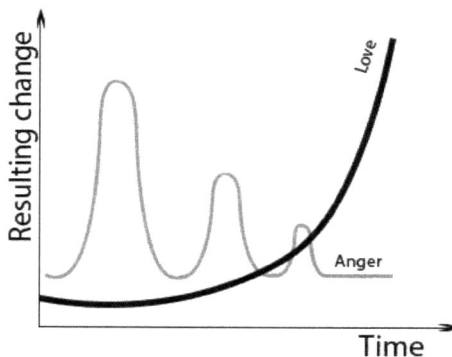

Figure 28. The long-term impact of harsh discipline and love. Despite impressive short term effects, the potency of severe punishment quickly evaporates. Love generates meaningful results only after building trust, the most crucial step.

---

[68] Spanking is non-injurious, open-handed hitting to modify child behavior.

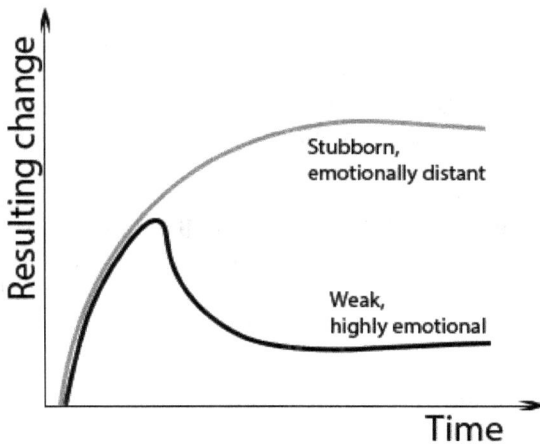

Figure 29. **Possible personality consequences of severe discipline over time.** Harsh punishments harden strong-willed children into inflexible, rigid personalities, while others break under its heavy burden. Learned helplessness is a well-known example.

and hide any vulnerabilities. We cannot denigrate a child and expect her to believe in herself. These insecure people can become ruthless, domineering personalities. The broken victims of these relationships often end up ensnared in a cycle of controlling and abusive behavior.

The Cycle of Abuse

An abusive relationship is a tumultuous journey characterized by agonizing lows and deceptive highs that isolate the victim from friends and family. While the objective of abuse is complete domination, it often begins with alluring promises to establish trust. Abusers are adept at discovering their victims' vulnerabilities by constantly pushing the limits of what their partners will tolerate.

The insecure abuser is suspicious of external contact and hints of resistance, which can provoke a confrontational reaction. They carefully choose moments when their victims are most defenseless and unobserved to mete out the pain but cause humiliation when there are witnesses. The depth of emotional wounds caused by this brutal behavior triggers the abuser's remorse, marked by conciliatory gestures and ostensible generosity as a form of 'atonement' for earlier cruelty.

In the quest for total control, the abuser needs ongoing validation, setting up a cyclical pattern of tension and relief, aggression and contrition. These oscillations deepen the emotional interdependence between the abuser and the victim, forging bonds that make it increasingly difficult for either party to exit the relationship.

This cycle of harm and subsequent reconciliation dominates the lives of those involved. Anticipating the next wave of abuse fosters a perpetual state of defensiveness in the victim. Over time, the victim becomes highly attuned to the abuser's cues and responds with preemptive submission to avert conflict. From an external viewpoint, this manipulative relationship can deceptively appear functional.

Abuse acts as a perverse coping mechanism for emotional and mental insecurity. Simplistic and biased rationalizations reveal the abuser's incomplete or skewed thinking. Abuse can bind people (or animals) to unimaginably abhorrent conditions and result in horrible personality changes for both sides of the mistreatment, exemplified by Festinger's theory.

Gaslighting: Gaslighting is a specialized form of abuse named after the 1944 film *Gaslight*, in which a husband crafts an utterly unpredictable environment to distort his wife's perception of reality. This manipulative tactic is not limited to intimate relationships; it is employed by narcissists, cult leaders, and tyrants to control larger groups or entire nations. Philosophers dating to Plato in 375 BC have identified the deeply corrupting nature of such manipulative control.

Preparing for Independence

The intense emotional bonds created in an abusive relationship act as shackles, which makes it exceedingly difficult to disentangle oneself from the situation. The first step toward achieving independence is envisioning a life free from abuse. Meditative practices or prayers can fortify internal resilience, integrity, self-confidence, and self-belief. The abuse victim's isolation makes seeking spiritual, organizational, and financial help challenging. Often, the victim is all alone on the arduous journey to freedom.

Understanding the Energy Dynamics of Abuse: Like other negative attitudes, abuse is an information radiation that targets its victim. As remembered from emotional fluorescence, this radiation creates contradiction (up-down spin pairing), which fosters emotional separation. While the impulse for some form of retaliation may be strong, vengeance only projects back the energy, inadvertently strengthening attachment.

It's vital to see the inherent fragility in the abuser and to respond with compassion and acceptance. This elevated perspective can be increasingly challenging because the abuser intuitively responds to understanding by accelerating the severity and frequency of attacks. This is where external emotional and social support becomes so crucial. Setting a realistic deadline for leaving the abusive situation can be immensely empowering.

Healing after Abuse: Exiting an abusive relationship doesn't necessarily sever the emotional ties between the abuser and victim; these can linger painfully, potentially for a lifetime. Is true liberation achievable? Acknowledging the abuser's insecurities and weaknesses can disarm their power and elevate your spiritual fortitude. In a transformative process analogous to a clam creating a pearl from an irritant, the agony of being a victim can be transmuted into a source of empowerment (Figure 7). Replacing fear and conflict with love and compassion provides the emotional strength for forgiveness. Breaking free from the vicious cycle of abuse, not just individually but also for the sake of future generations, deserves global attention.

Recognizing and acknowledging the existence of abuse is an essential first step in the healing process.

When the Belief is Crippled: Learned Helplessness

"People do not lack strength; they lack will." — Victor Hugo

Unlike saplings in a forest that can wait years dormant in the undergrowth for an opening in the canopy to grow into a tall and robust tree, the window of opportunity for physical and emotional development in animals is genetically determined. Early nutritional deficiencies stunt the growth of the body, but emotional neglect and abuse can similarly stifle mental maturation. The psychological phenomenon known as "learned helplessness" refers to intentionally undermining achievable tasks or the inability to escape adverse conditions (Figure 29). The resigned expressions of captive animals offer a poignant glimpse into their state of emotional defeat.

The Invisible Walls of Psychological Barriers: Our physical environment and societal frameworks often shape our mental outlook, creating invisible yet compelling psychological boundaries. These limitations can feel as real—if not more so—than tangible barriers made of rock and stone. The mere approach to these perceived boundaries can induce intense, hidden anxiety, making their transgression appear unattainable. One striking example is how a young elephant tethered by chains eventually gives up trying to escape, its spirit irrevocably crushed. Similarly, children who experience abuse often grow into adults who perpetuate the abuse cycle in some form. Though they might hide behind inventive excuses, the underlying issue is a deeply ingrained fear of facing challenges. As a result, the fear of success can ironically become a more potent deterrent than the fear of failure. Victor Hugo aptly said, "People do not lack strength; they lack will."

The Self-Fulfilling Prophecy of Learned Helplessness: Learned helplessness is the internalized conviction that failure is unavoidable. This belief can significantly limit one's potential for growth and achievement.

Shopping Addiction

> *"It is preoccupation with possessions, more than anything else, that prevents us from living freely and nobly."* — Bertrand Russell

It's almost inconceivable to think about life a century ago, a time devoid of basic amenities like electricity, running water, and household appliances. The quality of life for billions has improved dramatically, thanks to increasingly sophisticated technologies. Cell phones serve multiple purposes today; cars are more than just status symbols, and homes reflect our personal and professional standing. Our reliance on amenities, from comfortable beds to nourishing meals, ensures a stable, day-to-day existence supported by stores, restaurants, and cafes.

Personal Vulnerability: Aristotle once said, "All men by nature desire to know," highlighting our innate drive for novelty and the desire to derive meaning from it. We are attuned to social changes through social media, like never before. Instant comparison with peers via social media and the relentless news cycle exacerbate feelings of inadequacy, fueled further by targeted advertisements. We judge ourselves based on comparisons. Without meaning and purpose we feel lagging behind others.

To keep up others and the world, it is easy to seek material and financial pursuits. Such material acquisitions often leave us emotionally unsatisfied, creating a perpetual need for more, potentially spiraling into addictive behaviors. Shopping and other addictions serve an immediate, although painfully transient, gain of control to catch up to others, the community, and time.

The need to compete with the proverbial Johnsons makes us vulnerable to the sanitized dreams of advertisements. The probabilistic nature of unpredictable and intermittent rewards (Dabney et al., 2020) leads to their addictive nature (Rothenhoefer et al., 2021). Uncertainty, novelty, the sense of rarity, or special price triggers a highly addictive dopamine surge (Figure 30). Companies often offer discounts that require extra steps and require providing information to buy the item. The additional work makes the process even more rewarding.

The Illusion of Reward: Our vulnerability is further exploited by the tantalizing promises of advertising. The unpredictability of intermittent rewards has been proven to be addictive (Dabney et al., 2020; Rothenhoefer et al., 2021), triggering dopamine surges that encourage repeat behaviors

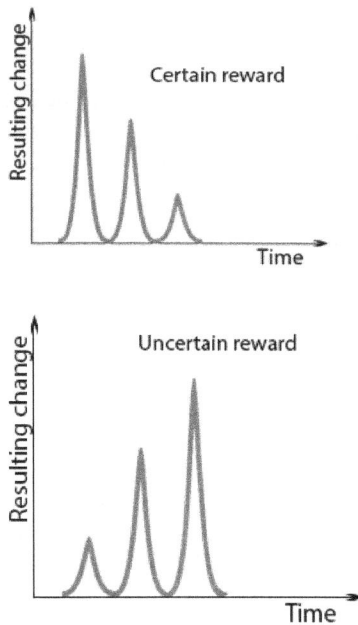

Figure 30. Dopamine level in response to reward. Dopamine output decreases after predictable (top) and increases following uncertain rewards (bottom). The excitement accompanying dopamine release can cause addictions. (Drawn after Starkweather and colleagues. *Neuron*, 2018.)

(Figure 30). Marketers tap into this by offering discounts that necessitate additional steps, creating the sense of a treasure hunt fueling its rewarding quality. However, it is often just wasteful spending.

The Pitfalls of Material Excess: When acquisitions don't fulfill genuine needs, they contribute to a growing sense of emptiness and disappointment. Such unnecessary items become costly clutter, relegated to forgotten corners of our living spaces. This accumulating disorder can induce a sense of latent unease, especially when pressed for time to find an item, only to face unmanageable clutter.

The Quest for Lasting Contentment: Ultimately, long-term satisfaction hinges on striking a balanced relationship between our needs and wants. Accumulating material goods may offer fleeting moments of happiness, but it seldom addresses our deeper emotional and psychological needs. Therefore, understanding the difference between what we need for a fulfilled life versus what we desire due to external pressures is crucial for lasting well-being.

Overcoming Shopping Addiction

Conquering a shopping addiction calls for disciplined effort, much like other forms of dependency. While chemical addictions often necessitate medical intervention and formal therapy—a subject beyond this text's scope—the shopping addiction approach involves a more psychological dimension.

Understanding the Root Causes: Our sense of self-worth and security influences our relationship with money and our willingness to spend it. While cutting off immediate access to funds can provide short-term relief, long-lasting change necessitates digging deeper to address the underlying dependency issues.

Strategies for Sustainable Change: (1) Redirecting Attention: Engaging in a meaningful and personally fulfilling activity can divert attention away from the urge to shop. This replaces the dopamine rush associated with shopping with a more constructive form of gratification. (2) Embrace Generosity: Donating excess possessions frees up physical space and provides emotional liberation. Redirecting focus from accumulating more to sharing and giving can encourage more mindful spending habits moving forward.

The Virtue of Patience: Success won't be achieved instantly; change is gradual. However, even small positive steps toward improving oneself can cultivate a sense of progress. This forward momentum alleviates feelings of restlessness and uncertainty. Owning up to past mistakes while actively shaping the present lays a foundation for a more empowered future.

The Closed Inner World of Depression

> *"In separateness lies the world's greatest misery."* — Buddha

Despite the immense technological advancements, navigating our social environment remains a tremendous challenge. Shockingly, almost half of Americans (a country where statistical data is available) resort to mood-altering, both legal and illegal substances. The global rates of mental illnesses, psychiatric diagnoses, and suicides are on the rise, and drug deaths account for more than half of all deaths of despair in the United States.[69] Depression, the primary cause of disability, affects approximately 4% of the world's population, according to the World Health Organization.

Physiological and Psychological Aspects: Depression's multidimensional nature is indicated by physiological and psychological symptoms, such

---

[69] Source: Human Mortality Database.

as worthlessness and guilt. Depression is a complex interplay of brain chemistry, where early-life stressors and even gut microbiome play crucial roles. For example, the possible role of inflammation in mood disorders and anxiety brings attention to treating the whole person rather than the disease.

The brainstem's quick perception serves our intuition. Compromised connections between the brainstem and other brain areas, particularly the frontal cortex, incapacitate long-term thinking. Without the future, attention is locked the past (Belmans et al., 2023), forming guilt and regret. The difficulty of remembering positive memories triggers an insidious cycle, intensifying the problem.

Modern life often feels fractured, without meaning and purpose. Workers often denied the big picture and forced into micromanaged roles. Lacking purpose themselves, parents fail to provide directions to their children.

Therefore, depression often originates early; childhood stress increases the susceptibility to depression later in life (Rousson et al., 2020). The powerless condition involves indifference to cognitive functions and thoughts. Rigidity is energy-expensive, degrading mental energy, exaggerating the vulnerable disposition. Telling someone with depression, "Just get over it", is like asking the desert for water. When the inner clock slows to a halt, it becomes as inescapable as a black hole. In a cruel twist, the weak mind carries the immense weight of its past.

The broader societal issues, including rapid technological advancements that, paradoxically, often leave people feeling disconnected, adversely affect mental health. The modern era in the developed world had witnessed the almost total subsuming of human life. *The loss of control or the lack of mental progress predisposes to depression.* Problems with low self-esteem and distorted social comparisons represent a sense of separation, triggering desperate attempts (bribes or abuse) to hold onto relationships. While our modern world offers many comforts, it often lacks the tightly-knit community bonds that can act as a buffer against mental health issues.

Treatment and Prevention: Healing does not come from the dark but from light. Openness shines understanding on our inadequacies, pains, and distress. Mindfulness-based interventions, cognitive-behavioral therapy, medications, and even newer approaches like neurofeedback (see Chapter Six) offer a range of treatment options. These could work best in combination and should be customized to suit the individual's unique needs.

While "healing by one's own hands" through art and handicrafts appears old-fashioned, these activities cultivate a sense of beauty, balance, and joy. Art-making boosts confidence, which can enhance resilience toward stress and depression. The creation of beauty, such as art and music, reformulates the mind. As a particular benefit, arts and crafts—knitting, music-making, and cooking—encourage positive social interactions by sharing resources and creations. The tangible achievement from our own two hands lends satisfaction. Even for trauma victims, these inspiring, meditative actions improve mood. Physical activities and sports are also helpful.

Grief: Depression causes separation, and inversely, isolation can trigger depression, becoming a spiritual death in a fully functioning body (González-Roldán et al., 2023). Chronic diseases, losing a loved one, or significant setbacks represent a lonely, outcast status. After a loss, it takes time to weave the irreversible changes into an integral part of our mental fabric. Still, excessive regret, guilt, or anger can inhibit recovery. The willingness to seek professional help might be the first decisive step toward healing. Acceptance builds hope, meaning, and wisdom.

Prevention and Resilience: It's essential not to overlook the aspect of prevention. Whether through fostering more supportive communities or individual actions like regular physical activity, small steps can sometimes prevent the onset or recurrence of depression. However, these are often easier said than done, particularly for those already in a depressive state.

Social connections form the glue connecting families, communities, and nations. Depression often starts with an almost irresistible and seductive attraction for separation from others. Separation is an exothermic process which robs our mental energies. Thus, replacing inward attention with meaningful social connections can overcome this destructive drive. For example, making a difference in other people's lives and the broader society overcomes separation and the energy-weak state. Deliberate actions toward small goals keep the mind open and engaged. Belief in a higher power can help renew hope in the possibility of healing.

Connection to Love: Similar to awe, depression can make time feel endless. However, unlike joy, which brings people together, depression creates a solitary and confined world with no apparent possibility for escape. Depression's isolating nature underlines the healing effect of social connections, particularly compassion and love. Breaking the urge of isolation by approaching a friend can initiate a non-judgmental relationship to lift us. Additionally, some of the tips outlined in "Success Boosters" may also aid in overcoming or preventing depression.

Social Fusion: Embracing a Paradigm of Collective Agency

In the preceding chapters, we embarked on a transformative journey, dissecting the fundamental nature of emotions and consciousness, and framing them as energy states within the brain. This perspective, apart from being an intriguing conceptual lens, bestows upon us a potential for mastery over our own lives and helping our loved ones doing the same. There are apparent ramifications of this understanding, particularly in the realms of education and parenting, where motivating young minds to flourish into their fullest capacities becomes a conceivable endeavor.

However, the voyage through these chapters also revealed stark, unsettling contrasts in how large corporations and organizations perceive and treat their human resources. Employees are deemed as mere disposable cogs in the giant operational wheel, in some way far more demeaning and dehumanizing situation than slavery, where slave's health was paramount. In contrast, government intervention is necessary to ensure basic safety and well-being for the workforce, exemplified by mandates on essential protective equipment within factories.

We now find ourselves at a pivotal crossroad. The societal view of women and minorities has progressed from a place of diminishment to a realm of deserved respect and dignity over the last century. In parallel, technological advancements, particularly the integration of AI in robotics, are redefining the dynamics on factory floors, empowering workers with unprecedented levels of responsibility and influence. This coupled with the demographic shift of a shrinking population, lays the groundwork for a probable elevation in organizational and financial reverence towards workers.

Ancient Athens, an important naval power, extensively used ships for combat. The enthusiasm and strength of the rowers, made up of common people, determined the maneuverability of the ships (Scheidel, 2017). This crucial dependence necessitated the City-state highly democratic social structure. Like ancient Athens, humanity's survival rests on the common people.

Therefore, ancient Athens illustrates a vibrant picture of democratic principles in action. The metaphor of Athenian rowers, whose fervor determined the maneuverability and thus, the fate of combat vessels, symbolizes the indispensable value and potential of collective agency. A thoughtful, responsible engagement enriches social cohesion, a fabric necessary for facing and overcoming the myriad challenges of our times. Social cohesion, where everyone feels respected inspires intelligent and responsible contribution for the common cause. Such social fusion is paramount to tackle the problems and challenges we face.

The impending AI revolution is poised to thrust every employee into a sphere of intelligent participation within industrial and corporate domains. This evolution echoes the urgency to tackle global dilemmas like climate change and waste accumulation, which can only be mitigated through a conscientiously caring citizenry. Human brain evolution endowed us with an ability to extract meaning and purpose, but this privilege was largely the privilege of artists, scientists, and philosophers. As the oppressed classes were shackled in a void of mental stagnation, the cruelty of oppression robbed the rulers of compassion, a necessary ingredient of creativity and meaning.

The voyage concludes by urging a renaissance of social structure, advocating for a unified, overarching common goal that resonates with every stratum of society. Through this, we not only acknowledge our intertwined destinies but also foster a culture of shared responsibility and mutual respect. Such a social metamorphosis could catalyze an era where each individual, unfettered by oppressive class systems, can contribute intelligently and meaningfully towards a harmonious, prosperous society.

The exploration in this book doesn't merely seek to provoke thought but aspires to ignite a movement towards acknowledging and embracing the profound potential of human emotions, consciousness, and collective agency.

Chapter 7

Success Boosters

"Where a technique tells you 'how' and a philosophy tells you 'what', a methodology will contain elements of both 'what' and 'how'."

— Peter Checkland

Societal and technological changes make adhering to our traditional political and social systems increasingly challenging. Environmental degradation, an aging population, and rapid technological advancements require us to reconsider our individual and societal ways of life. As the Industrial Revolution transformed the feudal system into capitalist structures, our technical progress necessitated a fundamental social and political fabric transformation. While change can be challenging and scary, the deepening societal conflicts indicate the need for structural change.

In modern cities, poverty is a glaring and unyielding reality, highlighting the frailty of our social structures and the importance of security. The constant threat of financial ruin might create the impression and belief that happiness is associated with monetary gains. A deeply ingrained belief—often rooted in childhood experiences—urges us to base our self-worth on physical appearance, social standing, and innate abilities. As a result, people often push themselves into emotionally draining jobs and unsustainable work hours to attain a veneer of contentment. Yet, these beliefs do not offer opportunities for personal growth. Unless we expand our horizons, we cannot move beyond our parents' limitations toward forming a meaning. With grit, determination, and risk-taking, we must continually remake ourselves via an active, optimistic lifestyle.

The Virtue of Continuous Self-Improvement: Drawing inspiration from Aristotle's philosophy, virtue emerges as a cornerstone for genuine human flourishing. Achieving long-term success demands more than temporary motivation; it requires an unwavering inner drive and a solid ethical foundation. This approach to life is not static but involves perpetual psychological growth. Undertaking the journey of self-improvement not only paves the way for career advancements but also enriches job satisfaction. However, this metamorphic process is not for the faint of heart; it's akin to summiting a mountain. Lenz's law in psychology shows that external circumstances and beliefs oppose rapid mental shifts. Therefore, gradual and consistent progress offers a more sustainable path.

Emotion Regulation as an Essential Life Skill

It requires a license to navigate complex machinery like cars. Yet, we operate the most complex organization in the universe, our brains, without understanding where emotions can take us. A deep understanding of emotion regulation can significantly elevate personal autonomy and self-confidence. This essential individual skill could be a valuable addition to educational curricula, empowering the next generation to manage their emotional landscapes effectively. As pioneers in this realm experience success, their stories can serve as a roadmap for broader societal adoption. The beauty of these self-improvement techniques lies in their universal applicability, irrespective of age, culture, or religious background.

Like traffic laws or multiplication tables, understanding emotion regulation can endow our social landscape with confidence and an Aristotelian virtue. Like traffic laws or multiplication tables, the rules of emotion regulation are independent of cultural, religious, or social background. Moreover, they can be adapted to any unforeseen situation.

A society of confident and moral individuals expects the same from its government. A cycle of mutual trust between governments and citizens can catalyze sweeping societal changes. The ripple effects of such transformation will manifest most tangibly in the lives of future generations, enabling them to unlock their intellectual potential within constructive community environments.

The impressive ability to save time and solve problems with AI offers incredible potential for achieving any goal. However, if we do not address the moral foundations of our society, the rise of AI could lead to a dangerous and explosive period plagued by crime and anarchy.

The Power of Incremental Change: Starting with acceptance and ending with love, the following twenty-nine points have essential, transforming power. Depending on individual circumstances and aspirations, adopting one or two practices at a time is the most effective.

In an ever-evolving society, standing still is equivalent to moving backward. Continuous growth and adaptation are aspirational goals and existential necessities for personal and collective well-being. Individually striving for better versions of ourselves contributes to a broader, more resilient, and equitable society. Our ultimate legacy will be the world we leave for future generations—a world shaped by our collective choices and individual transformations.

Unless we move forward, society is leaving us behind.

The Law of Orthogonality, Acceptance

"Understanding is the first step to acceptance, and only with acceptance can there be recovery." — J.K. Rowling

A large percentage of thoughts throughout the day are recurring. Like a familiar song, negative emotions like resentment and regret can stick in our minds like an infection. And negative feelings are never alone; like flies circling around a dead worm, they gather and multiply. Mistakes stir up memories, creating a mentally exhausting mixture of worries and remorse.

These deeply ingrained recurring patterns can populate our minds like an incessant melody. Self-doubt and uncertainty often manifest debilitating thoughts: "This always happens to me," "I can never succeed," or "I'm not smart enough." These pervasive patterns act as heavy chains, anchoring and draining our inspiration, irrespective of our innate talents and capabilities. When these automatic thought patterns take over our thoughts, we, in essence, cede control over our lives.

The automatic thought loops of self-doubt and uncertainty trigger thoughts such as "This always happens to me; I can never do it; I am not smart enough." Independent of personal talent and ability, their pessimism and powerlessness weigh down and drain inspiration. *When we lose control of our thoughts, we lose control of our lives.*

Lenz's law has shown that conscious attempts to dismiss worries or ruminations offer only a fleeting respite. Placing the blame outside of ourselves only strengthens the past's grip over our present. Fighting against our circumstances is analogous to pushing down a floating log or taming a storm, ultimately futile endeavors. Instead, we must recognize problems in our mental world as an indication of the need for personal growth.

Rather than focusing on exerting conscious control over our thoughts, we should aim to alter our underlying beliefs. Mental freedom is impossible without liberation from repetitive thought loops. Spiritual openness requires regularly getting rid of grudges and hurts. The discussion on the Pythagorean theorem has shown the transformative power of acceptance.

Consider the captured Roman soldier Mucius,[70] whose deliberate act of placing his hand in fire aptly demonstrated his fearlessness to his enemies. True confidence is not easy; it arises from an unwavering belief in one's capacity to face hardships, including the ultimate sacrifice. It's about expecting the best while remaining unattached to outcomes. By accepting our personal truths, we foster a unified sense of purpose and intellectual resilience, enabling us to devise integrated solutions grounded in the essence of our being.

Moreover, the practice of acceptance extends beyond ourselves to how we interact with others. It means setting aside ego, pride, and biases that prevent us from genuinely receiving help or compliments. In doing so, we replace antagonism with collaboration, trust, and kindness, epitomized eloquently by Mr. Rogers, who asserted that compassion paves the way to success (Figure 21).

In summary, the route to mental freedom lies in unshackling our minds from cyclical, negative thought patterns. This liberation comes not from suppression but from acceptance—of our circumstances, the necessity for personal growth, and our own and others' intrinsic worth. Embracing our vulnerabilities, including a broken heart, can be the first step toward profound healing.

Find Meaning

"It is not good for all your wishes to be fulfilled: through sickness you recognize the value of health, through evil the value of good, through hunger satisfaction, through exertion, the value of rest." — Heraclitus

Paul Gauguin's iconic painting, *Where Do We Come From? What Are We? Where Are We Going?* captures the essence of humanity's existential pondering. For much of human history, the grinding reality of poverty and survival left little room for such philosophical contemplation. The pursuits of our ancestors often stopped at the immediate necessities: marrying and raising a family. However, our modern, technologically advanced society provides us free time to seek answers to more profound questions and define our unique paths. Unfortunately, we lack guidance on how to find our purpose, and most of us are navigating uncharted territory. This has led to a global mental health crisis, with depression, anxiety, and suicide rates on the rise. It's essential to understand how emotions are regulated not only for our individual well-being but for the benefit of our social fabric.

Advancements in science and technology have elevated global living standards. Still, even the most prestigious jobs can feel hollow without

---

[70] In response to the threat of execution, Mucius stuck his hand directly in the burning fire to demonstrate that he did not fear death.

a sense of purpose. Our lives are increasingly dominated by busyness, a hectic schedule that leaves scant room for mental or spiritual nourishment. This addiction to the adrenaline rush of constant activity and information conceals an underlying spiritual emptiness, a vulnerability to fleeting pleasures, and a restless urge to constantly shift focus.

So, what defines success? If we liken life to a journey by train, what is its ultimate destination? We all seek a form of transcendence, be it through a higher power or a trust in the interconnectedness of the cosmos. Understanding our role in this grand tapestry lends our lives a sense of significance and fulfillment. The key to long-term well-being lies in embracing life as an integral element within this more expansive existence.

Although we have infinite potential, our responsibilities do not extend beyond what we can handle. By positively contributing to others and pursuing meaningful objectives, we gradually align our individual purposes with a more universal, cosmic order. We can only take advantage of life's immense possibilities by stepping out of our comfort zone. If you're comfortable, empathize with the pain of others; if you're suffering, allow yourself to be embraced by the world.

In essence, meaning is not a fixed point but a dynamic, ever-expanding vision that instills in us a sense of progress and forward momentum. While challenges abound, individual and collective growth opportunities have never been more substantial. Our existential quest for understanding should remain an integral part of our journey through life.

Authenticity

"Don't ask what the world needs. Ask what makes you come alive, and go do it. Because what the world needs is people who have come alive."
— Howard Thurman

We become susceptible to external influences and societal expectations when we lack personal meaning. This vulnerability often manifests as insecurity, shame, fear, and anger stemming from an intense desire for social validation. For instance, insecurity of one's abilities prompts a defensive stance to safeguard one's social standing. It leads to prioritizing others over themselves in a quest for approval, setting the stage for resentment if those favors are not reciprocated promptly.

Insecure people rely on their massive egos for protection. Emotional distortions can compromise our authenticity, skewing our words and actions until they misrepresent our true selves. The ego's excessive Temporal Mass forms emotional lensing manifested as biased speech, behavior, and personality. To free ourselves from ego-driven bias requires an honest self-examination and open dialogue with others to discover our core values and form our internal compass, which gives rise to an inherent feeling for healthy boundaries. These boundaries rooted in our subconscious form a protective mechanism providing courage to automatically ward off anything that could harm or threaten our purpose. Our core values become our guidance system: the openness to ask and the generosity to accept help.

Cultivating trust and gratitude fosters meaningful interactions, serving as the cornerstone for collective engagement. Emotional safety enables us to give selflessly without being attached to the outcome. In a space where nothing is expected in return, regret loses its foothold, allowing us to adapt swiftly to challenges and opportunities. Just as energy naturally moves from warmer to colder substances, emotionally calm individuals do not need to assert themselves through conflict; instead, they glean wisdom and understanding from each experience.

Because authentic individuals can trust and safeguard themselves, they enjoy the freedom that comes with emotional security. In contrast, by focusing on their ego, insecure people expose their core and exacerbate their emotional vulnerabilities. Authenticity, then, is about remaining true to our innermost selves.

Learn the Rules

"You have to learn the rules of the game. And then, you have to play better than anyone else." — Albert Einstein

In his seminal book *Outliers*, Malcolm Gladwell posits the 10,000-Hour Rule, suggesting mastery in any field requires approximately 10,000 hours of dedicated practice. This roughly equates to 20 work hours a week for a decade. A diligent learner can acquire comprehensive knowledge and expertise in a chosen subject during this extensive period.

However, becoming an expert entails more than accumulating knowledge and refining techniques. Technical mastery is the foundation, but a personal signature and originality require turning the chore of study into a confident play. The innovative genius requires ambition and constant restlessness to bring forth the future by destroying and recreating everything with sheer imagination.

Understanding the unwritten rules of a community or industry can provide helpful shortcuts. By studying the luminaries in their fields, aspirants can intuitively grasp underlying patterns and nuances. Experts can ask critical questions that uncover unexpected problems before their symptoms.

However, one peril of limited understanding is overconfidence. Mark Twain sagely warned, "It ain't what you don't know that gets you into trouble. It's what you know for sure that ain't so." Humility is the key to acknowledging one's errors, which opens the hard road from self-centered delusions toward progress.

Take the case of Hedy Lamarr, the Hollywood actress and producer who, during WWII, invented a groundbreaking radio guidance system. Despite her invaluable contributions, she received no recognition or compensation, mainly due to her outsider status and failure to navigate the industry's unwritten rules. This highlights that exceptional skill alone is insufficient; understanding an industry's political dynamics is also essential.

The Virtue of Solitude

Although learning from others is crucial, insights deepen into an expert understanding in solitude. In Tchaikovsky's words, such "fertile solitude" strengthens the spirit and develops an authentic voice and a love of the subject matter. A notable accomplishment resulting from seclusion is Einstein's theory of relativity, a monumental achievement, was primarily a product of such solitude. His words show appreciation for this working style: "The monotony and solitude of a quiet life stimulate the creative mind." Solitude does not mean social isolation! Engaging with friends, family, and peers is a necessary reality check for our efforts.

Setting Goals

"Vision is the spectacular that inspires us to carry out the mundane."
— Chris Widener

Youthful optimism uplifts the imagination of a positive future (Schubert et al., 2019). Goals solidify these paper tigers into the realities of the future. Studies have shown that goal orientation is crucial for success in humans, emotional animals, and robotic simulations (Wissner-Gross and Freer, 2013). Having a well-defined purpose channels our energy and reshapes our mindset to achieve these aspirations.

Setbacks and obstacles are to be expected. The journey toward worthwhile goals will undoubtedly involve uncharted territory, detours, and pitfalls. However, each stumble provides an opportunity to renew our commitment and grow. Just as professional sports teams invest in motivational coaches to help navigate disappointments, the most effective coaching comes from channeling our deepest desires into enthusiasm.

So, what constitutes success? If life were a train ride, where would your ticket take you? Selecting a goal that fuels your intrinsic motivation is crucial, acting as a catalyst for every thoughtful action, step, and decision you make along your journey.

Keep Tabs on Progress

Like glistening faraway mountains, hefty goals appear promising and approachable but prove challenging to conquer. Our confidence level determines our perceived distance to the target and its clarity. Accomplishments make our objective more approachable, but failures and disappointments move it hopelessly out of reach. The subjective nature of confidence makes mindset an essential element for success.

The strategy must be updated periodically because even slight changes can significantly impact it. A regular appraisal of the progress allows the mind to *take hold* of its objective via a higher perspective. Such intellectual upgrading is a necessary part of successful goal-getting.

Although focusing on the goal is essential, rigid, tight concentration inhibits self-correction. Maintaining steady progress is an art that requires continuous adjustments with dynamism and flexibility.

Steps of Achievement

The goal should be constant in the mind, guiding decisions representing the most direct yet feasible path toward realization. Each decisive step forward invigorates the brain's reward centers and bolsters confidence and belief in our abilities.

Significant goals can feel daunting, but they are often possible to break down into smaller, more manageable tasks. Beginning with a modest project, it becomes easier to build on it diligently, gradually extending our ambitions. Solving one problem (one tiny step) accomplished daily permits one to incrementally build self-assurance and expertise through small victories. The process can provide a stable foundation for more considerable success. As a high-speed boat charts a straight and steady course, sustained momentum builds unwavering confidence and minimizes doubt. A vivid, unshakable vision of the goal guides each step.

The Power of Compound Interest

In a big organization, career and leadership advance with individual confidence and skill. Still, in modern, technologically advanced societies, unprecedented access to information has democratized the personal ability to dream big and achieve even bigger. Like compound interest,[71] each significant step forward amplifies determination, further fueling progress. This creates a synergistic effect where our behavior and thoughts mutually reinforce each other, accumulating a momentum that propels us toward our goals (Figure 19). Although initial efforts may seem unremarkable, consistent and genuine work will eventually bear fruit, optimizing mental and physical resources for peak performance.

In conclusion, the power of a well-defined goal cannot be overstated; it serves as a compelling catalyst for success. The goal we set charts our course, influences our actions and ultimately shapes our destiny.

Anticipating a Change

"For all days prepare...when you are the anvil bear; when you are the hammer, strike." — Edwin Markham

In his bestselling book *Who Moved My Cheese?* (1999), Spencer Johnson delves into the psychology of decision-making in a world of constant change. He argues that to thrive and expand our horizons, we must not just react to change but also actively anticipate and drive it. This mindset demands a level of mental agility to adapt to shifting circumstances.

Acclimatization

Consider the parallel of mountain climbing, a pursuit that necessitates physical stamina and intellectual acumen. Here, unpredictable weather conditions bring a new significance to traits like determination and discipline. The climber's need to adjust to changing circumstances and acclimatize to high altitudes mirrors the adaptability required in our ever-changing world.

Building a Momentum: Educational systems build student confidence by designing curricula offering incremental achievement opportunities. Even small victories enrich our insight, whether understanding a complex joke or winning a game. These mental accomplishments can accumulate into a formidable momentum, as Newton's first law of motion in psychology shows. When students finally receive their diplomas, the ceremony represents a stamp of approval after years of dedication and effort. That sheet of paper, adorned with ornate lettering and backed by respected authorities, symbolizes a societal vote of confidence in the graduates, whether they're emerging as doctors, engineers, or educators.

---

[71] Compound interest is the result of reinvesting interest together with the principal sum.

Finding the Forward Path: Embarking on a lengthy journey often involves unexpected twists and turns, including the occasional U-turn. Headstrong pursuit of an objective without flexibility is rarely the best action. Advantageous opportunities can offer shortcuts, but the willingness to promptly correct course when required is a critical skill.

When decisive action is necessary, like switching projects or crossing metaphorical gaps', hesitation can be disastrous. The path of progress inevitably leads us through unfamiliar landscapes, opening the door to fears and uncertainties—a topic we'll explore in the next section.

In sum, adapting to change isn't just about surviving; it's about thriving in a world that won't stand still. From the mountaintop to the classroom, mastering the art of adjustment is critical in our complex, rapidly evolving environment.

Mastering Fear

"Everything you want is on the other side of fear." — George Addair

Our life pulses with our sense of security. When we feel confident, we make wise decisions, but difficulty or danger initiates anxiety and withdrawal. In crises, events outside of our control dominate our lives. Traumas and threats disturb the ebb and flow of the body's energy balance, reverberating to the muscle fibers and nerves.

Fear is both an asset and a liability; it's an evolutionary survival mechanism that can both save and paralyze us. While it serves as an alert system to potential dangers, it can also confine us to a state of inactivity and stagnation. When we cannot trust our environment, we either seek control and dominance or seek isolation, the "civilized' equivalent of the fight or flight response (Figure 10).

Conquer Fear

Fear narrows focus by illuminating only a small, dangerous section of reality. The danger urges steps to increase security but fear can never be eliminated this way because every protective action uncovers a new terrifying detail, causing paralyzing tightness. How can we escape fear? One way to confront fear is through proactive engagement. Facing it head-on can dispel the nebulous dread that often magnifies the perceived threat.

Conquering Fear through Action: Ultimately, we choose between action and inaction. Since fear thrives on immobility, any meaningful action, however small, can serve as the first step toward emancipation from fear's clutches. By taking charge, we challenge the validity of our fears and reclaim our sense of agency and freedom. No matter how daunting, taking a decisive first step dispels the clouds of fear and brings a newfound sense of empowerment. The experience of failure, though not enjoyable, is far more instructive than remaining paralyzed by the fear of uncertainty.

Simulating and Believing Courage: The rise of simulation technologies offers us the unique opportunity to confront our fears in a controlled environment. Even without technology, the mental exercise of visualizing a supportive environment can go a long way in mitigating anxiety. Buddha's quote, "Imagine that every person in the world is doing just the right things to help you", can be a potent psychological tool.

Accepting Our Mortality: Our existential fear is the fear of death. Embracing this inevitability can lead to fearlessness and immunity to defeat. In Hegel's words, "It is only through staking one's life that freedom is won." Stephen Hawking, who was diagnosed with a fatal illness at a young age yet committed himself fervently to scientific pursuit, exemplifies the difficulty and triumph of such a choice; he defied the odds and lived well beyond his doctors' prediction.

Impostor Syndrome

The impostor syndrome is a psychological pattern that reflects the fear of being discovered as a fraud. The psychology of the impostor syndrome can be explained by Lenz's law. It occurs when our beliefs remain out of sync with significant life changes, such as promotion. Our projected false reality will act back on our attitudes, corrupting them. While fear can bind us in this mental prison, the key to escaping it is preparation (internal) and external support. Practicing for challenging situations can dispel the cloud of inadequacy, and seeking emotional and psychological support can provide additional strength.

The Power of Beliefs

"You have to believe. Otherwise, it will never happen." — Neil Gaiman

Belief systems are the mental frameworks that shape our understanding of the world. Ultimately, they provide faith for a guiding principle to serve as a compass for our decisions and actions. Interestingly, these belief systems affect individual outcomes and have broader social implications.

Self-fulfilling Prophecies: We often unknowingly form behavior supportive of our expectations. In psychology, the concept of the self-fulfilling prophecy highlights the potent role that our beliefs play in shaping our realities. If you believe you'll fail, you're more likely to do so. On the other hand, positive expectations often bring forth a favorable outcome. Freud's observation of the long-lasting impact of early positive reinforcement—being a "mother's darling"—underscores the enduring nature of these formative beliefs. The triumphant feeling and confidence in success can bring actual success with it.

Forming a Belief: While we may take it for granted, gravity ensures a firm ground under our feet. Similarly, our families laid the early foundation for our mental well-being. Our trust and confidence in our future parallels the stability and predictability of our social surroundings. Without this security, we must form solid commitments for ourselves. We can turn an encouragement from a parent, teacher, aunt, or caregiver into a personal validation. Alternatively, mentally distancing ourselves from challenges can provide a better perspective.

Optimism as a Strategy: The ability to see the glass as 'half-full' is more than just a cliché; it represents a mindset. William James's notion that "optimism leads to power" isn't just feel-good rhetoric; it's a psychological truth. Pessimism can be a self-imposed limitation, while optimism, even if against all odds, allows us to stretch the boundaries of what is possible. This is not to undermine real challenges but to advocate for a mindset to better navigate them.

Belief as a Catalyst for Change: One could argue that all significant achievements in history started as an act of defiance against the status quo. We cannot achieve more than what is offered by our beliefs. Whether it was Archimedes with his lever or countless innovators and thinkers who challenged existing norms, their beliefs empowered them to change the world.

Adjusting our Sails: Beliefs are not static; they can be reshaped and reformed. By revisiting and revising your beliefs, you open the door to new possibilities and potentials. In essence, our beliefs set the boundaries of our world. Changing them doesn't just alter the view; it can expand the landscape entirely. Convictions can become more of a barrier than a bridge to move forward. Nevertheless, beliefs are not set in stone; changing them can redefine who we are and what we can achieve.

Introduce Order

> *"Every disordered soul is its own punishment."*
> — St. Augustine of Hippo

Instituting a predictable routine can effectively rehabilitate abused and neglected dogs. Establishing order as a pathway to happiness, well-being, and success is deeply rooted in philosophical traditions and modern psychology. The order provides scaffolding upon which we can build a stable life. Whether establishing healthy patterns for a child or establishing a new organization, a structure allows for stability and predictability, which offers a sense of security.

Prioritization: Structure can be limiting in nature, but it has a liberating aspect because it frees us to focus on what truly matters. These structural aspects allow us to navigate the complexities of life with greater ease. For example, introducing a systematic approach to tasks can reduce stress. People use The Eisenhower Matrix tool to categorize tasks into what's urgent and essential, not urgent but important, urgent but not necessary, and neither. Likewise, breaking a formidable project into smaller, more manageable tasks makes the challenge less intimidating and more achievable. This aligns well with the psychological concept of "chunking.[72]"

Mental Strength and Resilience: Experiences are made up of emotional building blocks. These "sensory primaries" improve insight into our internal motivations and the structure of our emotional intelligence. For example, distress can provoke the feeling of pressure, disappointment forms a sense of heaviness, shame might parallel a sense of falling, and anxiety triggers an internal tightness. This evaluation is beneficial when we are at a loss in a situation. Confusion only makes finding solutions more difficult. Like pushing the gas pedal in a car is stuck in the mud, hard work does not correct the wrong approach. Asking the right questions can discover the root of the problem, making solutions possible.

Mental Order: Tambini and Davachi (2019) pointed out the cognitive tendency to form expectations for the future based on past experiences. This can lead to bias in how we perceive and process new information. However, even without a specific plan, our minds naturally create action steps toward achieving clear objectives. Therefore, good outcomes and straightforward solutions depend on mental order, particularly for larger projects.

Discipline and order are particularly essential in accomplishing big projects. Recognizing our shortcomings is the first step to rain in our handicaps. Empty entertainments and purposeless actions are like empty calories, leaving us scattered. Mentally nourishing experiences of beauty, learning, and joy can make us more resilient and able to handle life's complexities.

Financial Stability: Our ability to have order in our life also includes finances. Economic security can provide the room to breathe, make choices, and

---

[72] Chunking is a method of facilitating short-term memory by grouping individual pieces of information into larger, more familiar (and therefore, more easily remembered) groups.

navigate life's ups and downs. Because possessions and attachments can also weigh us down, the economic 'order' can directly affect psychological well-being and our ability to manage change. Accumulating a safety net by living beneath our means ensures long-term financial freedom.

Although money cannot buy happiness, higher-income people are generally more satisfied because financial safety ensures a higher quality of life. For example, assistants and gadgets can free them from mindless tasks, allowing a broader vision. With the advent of AI, these services are increasingly available for everyone, permitting democratization in planning and managing our lives.

Not extraordinary talent nor exceptional determination, but mental order is the most crucial ingredient for steady progress. Next, we turn to habits essential in creating order or structure.

Forming Effective Habits

"If you are going to achieve excellence in big things, you develop the habit in little matters. Excellence is not an exception; it is a prevailing attitude." — Colin Powell

According to research conducted at Duke University, about 40 percent of our daily actions are habitual (Neal et al., 2011). Charles Duhigg's 2012 *New York Times* bestseller, *The Power of Habit*, introduces the robust brain regions behind habits driven by the reward molecule dopamine and other neurotransmitters. Conscious action requires emotionally motivated effort to produce. However, repetition increases synaptic connection strength, reducing cognitive effort and emotional involvement. Therefore, repeated action sequences eventually become automatic and can be performed efficiently without deliberate focus, forming a firm foundation over daily life. Although habits can be performed without conscious concentration, they can lead to mental fatigue after a more prolonged engagement.

Feedback loops, such as a reward after a typical behavior, form the foundation of routines. As a result, regular practice maintains habits and provides a structure for our lives. However, transitions to habitual control lead to inflexible regulation, which can be a disadvantage during sudden changes.

Habits' reliable stability indicates that they are part of the temporal field. Their temporal nature emerges from projecting past behavioral control into future action sequences. However, we must differentiate between good and bad habits. Immediate reward quickly solidifies bad habits, whereas good habits require conscious effort to develop and maintain. For that reason, good habits are easy to lose; if the practice ceases, it takes renewed commitment to resume it.

Bad Habits: Analyzing the origin of a bad habit can offer a path to gain control. For example, journaling can help uncover their connection to our reward system. However, eliminating a bad habit is difficult even after understanding its roots because of the link to dopamine release. In addition, the limbic system in the brain activates the fight-flight-or-freeze responses, which lead back to the habit. As a consequence, it is easier to replace an unwanted habit with a more desirable one.

Even so, it takes concentrated, painstaking practice over weeks or longer to achieve results due to the difficulty of changing our automatisms. However, as change starts, it increases confidence, which inspires further improvement. Replacing just one set of behavioral patterns can have long-term consequences. Slow, gradual changes redirect cravings and *reprogram* our lives.

Practice turns imperfections into strengths, which leads us to persistence.

Persistence

"Many of life's failures are people who did not realize how close they were to success when they gave up." — Thomas A. Edison

The famous 1968 Stanford marshmallow test[73] is an experiment testing preschool children's willingness to wait for a reward. It showed that patient children had higher SAT scores and greater economic security in adulthood. Recent work confirms the results for children whose mothers had not completed college (Watts et al., 2018) but cautions from broader generalization of the findings.

The uncertainty over the above results indicates that our early life does not define our accomplishments: projects we are passionate about can inspire remarkable persistence. Perseverance is the inherent ability to regulate emotions, thoughts, and behavior (self-control). However, our mindset is influenced by our physical and psychological needs. Maintaining persistence and determination when we are frustrated or tired is hard. Even minor inconveniences, like hunger or a full bladder, can quickly diminish our patience. The following section introduces the role of persistence in the three stages of task performance.

Start: The courage and mental investment needed to start a project keep some people from moving beyond their dreams. Their ambitions for the future never move beyond the planning stage.

---

[73] Children were given a tasty snack, such as a marshmallow, and told that if they could wait for a short while before eating it then they will get an extra snack as a reward.

As remembered, Newton's first law for the mind shows the importance of acquiring mental momentum. Learning a chosen field's basics can be a good foundation for long-term progress. Persistence helps to keep going during these first steps when efforts are not always rewarding. Progressing too quickly without proper understanding at this stage denies the chance to embrace the project thoroughly. Ignoring the basics can lead to early burnout or exhaustion.

Therefore, there is a difference between doing things quickly and right. Like a little creek forming a Grand Canyon, initiating a big project with a tentative, slow, steady investigation can help find the optimal path. Similar to the opening scene of a stage production, where the plot is introduced with a foresight of the conclusion, starting a new project requires envisioning its completion. Proper care and consideration at this stage may include noticing discrepancies in the accepted professional dogma. Asking the right questions can offer the best path toward progress. By adjusting our ambitions gradually, we can pick up momentum as our experience and knowledge base increase.

Middle: Like a tightrope walker who keeps his eyes fixed on his destination, unwavering belief in the project's success drives progress. It is vital to maintain a steady focus. If we concentrate too narrowly, we get lost in the details. If we focus too far, we lose sight of individual steps.

The difficulty of the middle section is the grinding nature of the process. It is crucial to realistically evaluate our progress, and we must persist in the face of discouragement and naysayers. Healthy skepticism triggers honest examination, but doubts are defeating. A sustained belief can secure progress and ensure the energy to start over if external obstacles or mistakes block our progress.

Finish: The care and consideration needed to finish a project differs significantly from the earlier steps. Close to completion, impatience can urge us to rush things toward completion without proper care. The other mistake is constantly upgrading expectations, which can indefinitely delay the project's completion. Striking the right balance requires a realistic view of our project and an understanding of the external environment or our chosen field. Comparing the qualities of our product with the existing competition can help us navigate the project's conclusion.

We must gradually adjust our approach with discipline and not lose our determination at the first sign of triumph. Reaching the last mile in the marathon, adding the final brick to the house, and reviewing the presentation one more time moves the project to the home stretch.

We should not settle for where we are today; the pace of technological progress makes it increasingly essential to continually further our education and skill set. This takes us to responsibility.

Responsibility: Promises and Commitments

"The promises of yesterday are the taxes of today."
— William Lyon MacKenzie

We live in a society with norms and expectations. Trust is the foundation of social existence. We trust that cars stop at the red light, the legal system serves the victim, and firefighters and paramedics save people. How do we form a trust? According to a 2018 study, trustworthy people consider the interests of others besides their own. Respect for others makes it natural to act honestly and accountably in all areas of life (Levine et al., 2018).

Consideration for others is also connected to trust in others. As individuals, we can inspire trust and confidence in those around us by embodying integrity, honesty, reliability, and loyalty. These qualities have the tendency to fill our environments and spread to others. Newton's third law, "For every action, there is an equal and opposite reaction," implies the reciprocal nature of trust in relationships.

Promises: Promises are integral to job descriptions, appointments, and marriage contracts. Nevertheless, decency moves obligations beyond written agreements. Keeping commitments builds trust, which goes a long way in creating a positive environment.

Insecure people rush to make promises because they are governed by short-term motivations. Their weakness makes it difficult to say no, inspiring them to accommodate everyone. However, their mind changes with the prevailing social winds, making it easy to forget their commitments. Unfortunately, broken agreements erode trust and exaggerate uncertainty.

Commitments: Being accountable is essential in developing confidence and self-esteem. Living up to our families' and communities' standards internalizes those expectations until they become second nature. However, accommodating every request mindlessly, without understanding or verification, is a sign of weakness. Without the strength to say no, we compromise our secure boundaries, leading to remorse in the long run.

We can create a positive environment by committing to our principles and beliefs, speaking the truth without deceit, honoring our commitments, and standing by our loved ones and friends. Being impeccable with our words intersects with and influences our ability to maintain mental order.

The Power of Attitude

"When you take care of your effort and intention, with a little bit of luck, outcome will follow." — Amit Sood

The role of attitude cannot be overstated, as it significantly shapes our perception, awareness, and behavior. Like a powerful magnet, personal disposition guides perception, cognition, and behaviors by coloring every question and perception in our minds. While happy people may find encouragement when fearful people see danger, a positive attitude can open the world to us, making everything seem possible.

Culture and traditions are critical in determining acceptable behaviors in families or countries. Collective attitudes shape societal norms and expectations, particularly in the age of social media, where opinions can change rapidly. Peer pressure can modify views about the stock market, housing, or new technologies. Therefore, national experience, as seen with atomic power, for example, can dramatically alter what is socially accepted or attractive.

Choosing our Focus

Our focus is crucial in determining the details of our observations, much like the microscope's magnification. Adjusting our attention can help us see situations in new and enlightening ways, just as a microscope reveals layers upon layers.

Because changing the microscope's depth perception takes an effort, it takes a conscious effort to modify our attention. For example, mistreatment, abuse distorts vision, causing distrust or cynicism. Such bias misconstrues reality to fuel aggression, racism, cults, and repressive regimes, highlighting the importance of having a clear and unbiased perspective. Forgiveness is difficult but overcomes accumulated tightness by releasing built-up hate, resentment, or anger. Reformulating our narrative as a doer and success story can help rejuvenate our well-being.

Taking responsibility for mistakes by correcting and compensating for misdeeds is essential because it eliminates intense guilt and shame. Instead, self-forgiveness clears the mind of the weight of the past, opening the path to personal growth and positive motivation, enhancing mental order and persistence. Genuine self-reflection is an integral part of positive transformation, which tends to reverberate positively in our environment, affecting those we interact with regularly.

Because attention acts as a magnet for our attention, following every lead that competes for our attention fractures our ability to focus. Learning to give up our wants can be a liberating experience that clarifies our focus on our needs, allowing us to achieve our goals and live a fulfilling life.

The Power of Learning

"Education is the passport to the future." — Malcolm X

Entitlements and privileged positions of kings, emperors, princes, and noblemen were often attributed to God. Feudalism during the Middle Ages had solidified into an inflexible hierarchy. Individual interactions were dictated by strict legal, economic, and military rules, making human interaction across class differences virtually impossible. However, the Industrial Revolution disrupted this social order, creating an opportunity for those outside the aristocracy.

The hierarchical structure of titles yielded to the power of finance, as money and connections drove political and social advancement in this new era. In the twentieth century, modern economies moved toward a knowledge-based structure, with information, expertise, and intellectual capital forming its central pillar. Education and learning became crucial to upward mobility.

As society changes ever faster, social media and the internet have become pivotal in shaping social attitudes and mobility. Concurrently, there is a growing expectation for respect of the individual, reflected in film, politics, arts, sports, and science. Increasingly, the democratization of information is once again transforming social mobility. Lifelong learning has the potential to level the playing field and act as a great social equalizer. Personal growth and transformation can be achieved by integrating learned knowledge and experience, ultimately determining one's social status.

Learning and the Mind

Learning goes beyond just qualifying for a job or gaining new skills. It expands the mind, promotes spiritual balance, and cultivates patience. Hard work, dedication, and persistence lead to long-term fulfillment and increased life expectancy (Hummer and Hernandez, 2013). Education helps to clear mental clutter, promoting positive psychological states and leading to better financial and health decisions. It also inspires individuals to make a positive impact in their communities.

Education is a conscious process that stretches the mind through two stages. The first stage involves gathering, absorbing, and conceptualizing material, which can be stressful and confusing. The second stage integrates the new information through a liberating and empowering process. Active learning involves engaging students in hands-on activities that require them to think critically about their actions (Bonwell, 1991). This approach is particularly effective in the digital age, where abundant online information makes lexical information less critical. The internet's security and accessibility foster openness and encourage knowledge-sharing.

Taking breaks between practice sessions can help improve memory consolidation of skills (Buch et al., 2021). During these rest intervals, the brain replays faster versions of the patterns from the practice activity, which boosts learning. This replay is specific to the trained sequence and can predict the amount of skill consolidation.

Mimicking (Copying)

Copying is the first learning method for children. In a positive environment,[74] it is often also the most efficient way to acquire knowledge and new skills for adults.[75]

Observing and imitating superior personal and professional performance is a well-traversed path toward professional success and well-being. In a trustworthy environment, people may unconsciously imitate each other's mannerisms and posture, which can build social capital and spread attitudes and opinions through social networks. For example, Sentiment Contagion expresses the spreading of views and preferences among social network members, whose effects are not always positive. For example, stock market bubbles might originate in the shared optimistic beliefs of enthusiastic customers, which are disseminated and amplified through social interaction.

Copying within peer groups and organizations permits employees to absorb the culture and unwritten rules of their workplace through mentoring and imitation. As newcomers gradually build confidence, it produces a shared culture, influencing personal and corporate decisions.

Rote Learning

Learning through repetition, known as rote learning, is a standard method for mastering challenging material. Songbirds, such as zebra finches, use this technique to learn complex songs by precisely tuning connections between neurons. This stepwise approach is also used to memorize mathematical tables. Consistent exposure to the material can lead to muscle memory and automatic action sequences. Rote learning can provide a foundation for technical mastery in theater, dance, music, art, drawing, and language.

Beyond teaching a new skill, rote learning is a meditative practice that potentiates positive psychological states, attention, and patience. The technique clears the mind from wayward thoughts, generating a fresh,

---

[75] Cook et al., 2012.

empowering, and energetic feeling. Gradually increasing the repetitive routine duration over time can significantly boost mental energy.

Operant Conditioning

The concept of operant conditioning links a person's actions to the direct consequences that follow. These consequences, whether positive reinforcements or negative punishments, will determine whether the behavior will likely be repeated. In the corporate world, doing a job well can advance a career. At the same time, poor performance can lead to demotion or even termination. Various motivations can impact the effectiveness of operant conditioning. For example, this learning method may not be effective for self-motivated individuals who do not respond well to threats.

Lifelong Learning

While a college degree is often a requirement for specific jobs, it is not essential for leading a successful life. The demand for diverse skills in technology and social services is at an all-time high, and the need for continued learning has never been greater. Inversely, obtaining a college degree does not mean that learning stops there. The flexibility of intellect makes its improvement possible, an increasingly necessary requirement in our rapidly evolving society.

The Role of Advice

> *"Never react emotionally to criticism. Analyze yourself to determine whether it is justified. If it is, correct yourself. Otherwise, go on about your business."* — Norman Vincent Peale

Critiques evaluate work in the arts, sciences, and other professions for the recipient's benefit. The distinction between constructive feedback and judgment is crucial. While constructive feedback provides encouragement and support, judgment has a negative connotation. While the former is meant to guide and better the individual, the latter often arises from biases or vested interests, which are often betrayed by their tone beyond its oppressive message.

An environment thriving on judgment will not foster creativity, collaboration, or growth. The degree of personal attacks and reproach within an organization is inversely proportional to its flexibility and progressive drive. Resilient organizations are highly dynamic and show lesser polarization (Klimek et al., 2019). They champion constructive feedback and view it as an avenue for growth rather than a threat.

Response to advice is a test of the recipient, indicating his inner strength. Below, you find three possible reactions to criticism—the list represents the increasing independence of the recipient, showing learning and confidence stages.

1. Insecurity

Our opinions are rooted in our morals, values, and principles, representing an expectation of how we think the world should work. Pronounced sensitivity to what others say signals insecurity. This narrow, defensive viewpoint can lead to a restless striving for perfection, degenerating into a tendency to disappoint others. For an insecure person, even compliments can question his firmly-held ideas. For example, praising children with low self-esteem predisposes them to shame (Brummelman et al., 2014). Playing the victim or demeaning the self builds walls across every road that might lead to success.

2. Accepting Advice

As individuals mature emotionally and cognitively, they recognize the value of external perspectives. Advice is a much-needed second perspective empowering us to see ourselves through the eyes of others. Although it is vital to discern constructive advice from mere judgment, we grow every time we respond graciously to criticism. Even unfair decisions can deepen understanding and sharpen insight. Listening and considering input from others is a sign of respect and a powerful tool to learn, grow, and develop. In the words of Dale Turner, "The error of the past is the wisdom and success of the future." We develop inner strength and expand our limitations when we correct our mistakes.

3. Inner Strength

Understanding ourselves and our dreams inspires firm beliefs and self-respect in handling feedback. Because individuals at this stage profoundly appreciate their values and beliefs, they are open to feedback. They can stand firm in their convictions while being receptive to new perspectives. Their resilience is not based on avoiding criticism but on the ability to sift through it and take what's beneficial.

Authentic people form healthy boundaries, which provide independence against the most severe opposition and attack. They value the truth as a tool for self-betterment.

Broader Implications: Embracing feedback and using it as a tool for growth is vital for personal development and the betterment of communities and organizations. How feedback is given and received can shape organizational culture, leadership styles, and interpersonal dynamics. Moreover, understanding these stages of response can be instrumental for educators, managers, and leaders to tailor their feedback approach to various individuals, ensuring it is both practical and nurturing.

Decisions and Mistakes

"A person who never made a mistake never tried anything new."
— Albert Einstein

"The new basic principle is that in order to learn to avoid making mistakes, we must learn from our mistakes. To cover up mistakes is, therefore, the greatest intellectual sin." — Karl Popper

Arthur Miller's 1949 play, *Death of a Salesman*, is about the wrong decisions reverberating in our lives and the lives of our loved ones. Because our actions have inertia, mistakes must be corrected to avert their gathering harmful momentum. The inability to face and correct errors accumulate into a tsunami that engulfs and buries our dreams and future. Their menace can survive in the hidden corners of the soul for generations, exploding and destroying hopes at the most inopportune time.

Course Correction: Much like navigating a physical journey, our life also demands periodic introspection and adjustments. Getting sidetracked from our goals indicates the need to reexamine our objectives and correct our way. Rephrasing or forming the problem in a question form can quickly prompt a meaningful solution.

Embracing Mistakes

Mistakes are a necessary part of learning because they provide growth opportunities. Therefore, being unwilling to acknowledge and learn from our mistakes is detrimental.

The ability to rebound from mistakes is essential to resilience and vitality. Admitting wrong and apologizing is the best form of ownership, forcing us to move out of our comfort zone. Being at peace with our core values permits us to listen to our best judgment, naturally making us less error-prone.

Failure as a Catalyst

Failure is a painfully apparent persuasive argument for change. First, we must recognize the blunder before we can overcome it. Transformation is not easy; it requires work, effort, and time. Correcting failings is the best reinforcement of our goals. Doing the right thing requires courage but generates courage.

The Power of Intuition

In an age of information overload, intuition's role has become increasingly important. While data and facts are crucial, the human mind's ability to intuitively discern the best course of action, especially in complex scenarios, remains unparalleled. People with a strong sense of purpose trust their intuition and find greater satisfaction.

As discussed earlier, intuitive decisions follow psychological comfort, which can slice through conflicting information. Although insight improves with the decision maker's experience and confidence, the more complicated the problems are, the greater the advantage intuition provides.

Emotional Intelligence: An understanding of oneself and the emotions of others can lead to better, more empathetic decisions. Therefore, emotional intelligence and a clear purpose can guide us more truly than mere facts alone.

Taking a Break

> *"Every now and then go away, have a little relaxation, for when you come back to your work your judgment will be surer."*
> — Leonardo da Vinci

Biological, physiological needs, circadian rhythms, and psychological requirements rely on the brain's superior energy turnover. The quality of our lives depends on the pace of our day.

The Detriments of Chronic Stress: Our biological and psychological needs are intertwined. Modern lifestyles often subject individuals to sustained levels of stress. Stress and anxiety can worsen conditions such as diabetes and heart disease and accelerate aging.

Proper rest is crucial for maintaining good health. Creating gaps in our engagement calendar frees time for reflection and relaxation. Rest is a temporal oasis that refreshes the mind, increases resilience, and renews determination. Rest allows appreciation of the small things in life, their tiny, delicious pleasures, moment-to-moment inspiration, and fulfillment.

Resting is often associated with laziness but involves engaging in activities that rejuvenate the soul. Active rest flushes out the dust of everyday life. It can include attending a concert, visiting a gallery, walking in nature, or reading a meaningful book. Even mundane tasks such as cooking, cleaning, or organizing can be relaxing. For example, socializing with supportive friends and family or experiencing humor can boost confidence instantly. Likewise, helping others can improve our mental state and increase goodwill.

As our biological and psychological needs are interconnected, it is vital to prioritize rest and relaxation in our daily lives. More discussion on rest can be found in "Learning and the Role of Sleep".

Healthy Living and Exercise

"You're only one workout away from a good mood." — Unknown

As human beings, our physical bodies are composed of the dirt under our feet. At the same time, our minds carry the reflection of the universe or God. Recognizing the boundless potential within our consciousness must inspire us to prioritize our psychological and physical well-being. Therefore, taking care of oneself is more than immediate comfort; it also serves self-respect and ambition.

We expect our bodies to function for about a century. Investing time to learn the needs and potentials of our bodies is a worthwhile endeavor.

Bodily and Mental Benefits

Physical and Mental Advantages: Focusing on the future is a testament to valuing oneself. Self-improvement serves as a stable framework for meeting our health needs and objectives. A disciplined diet that caters to both physical and mental health is crucial. For example, eliminating a food group entirely is more effective than limiting specific foods. Alternatively, fasting for twelve hours or more limits calorie intake and switches the cell metabolism from sugar-based fuel to a slower metabolic process from fat. In turn, hunger helps regulate blood sugar, improves stress resistance, and suppresses inflammation. Fasting can also stimulate autophagy, a cellular 'clean-up' process, and enhance metabolic efficiency. However, it's important to note that while intermittent fasting can benefit many, it may not suit everyone, and it's advisable to consult with a health professional before making significant dietary changes.

Exercise—The Vital Habit: Exercise keeps the body strong and healthy, and the muscles send signals to the brain, positively influencing mental health. For example, while weight-bearing movements increase muscle mass, they also enhance the biochemical foundation of the nervous system. Exercise triggers the creation of new neurons and releases feel-good hormones. It improves physical intellectual performance, memory, and life functionality.

Regular exercise protects against numerous diseases and helps maintain cognitive function as we age. Exercise is the only proven way to stop or reverse physical and executive decline during aging, making it the ultimate fountain of youth. It extends the energy and agility of youth. So, just do it.

Letting Go of Control

"You must learn to let go. Release the stress. You were never in control anyway." — Steve Maraboli

We all want to protect our loved ones. Still, worries can overpower our common sense, pushing us toward closer management at any price. When we feel insecure, we hold our loved ones closer. Control is born out of good intentions but goes awry because of distrust of the environment or our loved ones. Strong attachments to people, things, or specific outcomes limit options and encourage an unhealthy and abusive need for control. The Buddha said, "The root of suffering is attachment." The need to always be right; the fear of being wrong makes us prone to mistakes.

Power of Surrender: Although letting go of control has the chance of loss, it offers profound relief and inspires learning. Surrender allows the natural tendencies of things to take shape and materialize. It suggests that sometimes, instead of forceful action, what's needed is a step back, a pause, a willingness to see what unfolds. This surrender isn't about inaction but trusting in the process and recognizing that we can't always dictate outcomes. There's a paradox at the heart of control; only by letting go can we have better relationships.

Organizations

Cultural formations and rituals have a strong presence. History, people, customs, and knowledge promote belonging, cohesion, and purpose. New ideas are resisted initially but, once accepted, treated like sacred knowledge. The importance of cultural evolution can't be understated. When organizations hold onto outdated models or practices due to fear of change, they risk stagnation or even decline.

Abandoning outdated expectations helps a project develop its organic direction. Relinquishing close control encourages group ownership, encouraging broader participation and contribution independent of position. Community goals with shared support generate a stronger momentum, which can sweep everyone with it.

Adaptability is recognizing when to shift direction toward innovation. For example, Netflix, a mail-order DVD company, evolved into an online server; Lego, a toy brick maker, achieved mega-success from collaborations with Harry Potter and Star Wars. In contrast, Kodak was synonymous with photography but had to declare bankruptcy in 2013. While Netflix and Lego evolved and adapted to changes in the market, Kodak clung to its film-based model, missing out on the digital photography revolution.

Control is often a journey inward. Recognizing our fears and insecurities can shift our behaviors and approaches toward a more active stance. The balance between control and flexibility can determine long-term sustainability and growth in both personal and organizational contexts.

Meditation, Prayer

"When meditation is mastered, the mind is unwavering like the flame of a candle in a windless place." — Bhagavad Gita

Our awareness has a rhythm undergirded by our psychology. Our mental power allows us to rise above minor annoyances or adversities. Therefore, psychological stability is not the lack of adverse experiences but eliminating their long-term consequences through establishing a higher perspective.

Meditation and prayer are prime examples of guiding one's thoughts. Eastern religious practitioners and sages perfected meditation into a centering yet relaxing practice. Meditation achieves a higher perspective by directing attention toward the present moment. Slowing the 'momentum' of our thinking encourages mental coherence, improving sleep, cognitive flexibility, and learning (Goldberg et al., 2018). Prayer is a type of meditation with a focus on a divine being.

Meditation and prayer are only as potent as the focus you devote to them. In addition, like other psychological treatments, they can induce transient negative impacts (Britton et al., 2021), with religious participants and females more immune to the unpleasant side effects. Therefore, to be successful, these practices need sufficient mental grounding and patience in the first place. Otherwise, they must be introduced slowly with greater awareness of thoughts and intentions.

The type of meditation should be appropriate for our personality, life situation, and specific problems. Quote Meditation and Prayer are introduced below; many other spiritual relaxation methods can be found online. Nevertheless, it is feasible to invent our own personalized way.

An Outside Viewpoint: Meditation is a tool to eliminate the information-loaded viewpoint. Meditation emphasizes independence from the world and replaces self-centered, stressful states with emotional neutrality, equanimity, and wisdom. The privilege of observing the self as if from the outside creates distance for a broader perspective. How do emotions feel like? Noticing the physical manifestations of feelings, such as their pressure, expansion, or tightness, provides an empowering insight, allowing changing their meaning.

Change the Meaning by Reprogramming the Brain: Like a heat pump that can switch between cooling in summer and heating in winter, perceptions can fuel positive or negative motivation. While positive affirmations accumulate mental power, negative ones lose mental power, and their effects can reverberate through time into our present. How can we transform the adverse effects of past experiences?

As shown in earlier chapters, beliefs update suddenly, discretely in the mind. A good example is riding a backward bicycle. On this type of bike, the front tire turns in the opposite direction of the handlebars, making it necessary to unlearn traditional riding. It takes several months of diligent practice to reprogram the neural and muscle programming within the brain. While the synaptic connections may modify slowly, the brain suddenly adapts to the skill of backward biking in an empowering moment. Therefore, we can leave the pain of the past behind if we methodically rewire our brains. In other words, to change the mind, we must reprogram the brain.

Cognitive Behavioral Therapy: Cognitive Behavioral Therapy (CBT) is a standard treatment for anxiety and depression. It entails invalidating, minimizing, or controlling unwanted thoughts. However, we have seen that emotions are energy states. The difficulty of thought suppression shows the impossibility of conscious control over our emotions. Therefore, changing the brain's programming is necessary to *update* beliefs. It is based on the understanding that the observer mind can only hold one vision or conviction of any subject at any time. Like changing the vision of the Necker cube, adopting a new view forces us to abandon the faulty opinion simultaneously. After the belief changes, the emotions and thoughts necessarily follow.

Quote Meditation

The transformative power of quote meditations lies in the consistent dedication to understanding and internalizing the wisdom within the quote. A quote distills a complex problem into its essential nugget with authority, and meditation internalizes its meaning. Like a water leak, starting as a drip but turning into a steady flow, the practice slowly turns into a sudden and discrete mental shift. At the moment of transformation, a sudden mental expansion signals the switch to a new belief. Like learning the backward bicycle, significant changes require regular practice over weeks or even months that institutes a switch between two mental programs.

Nevertheless, positive effects are noticeable much sooner, as the training gradually renews and upgrades our belief system. The recurring practice attacks and weakens tightly held and restrictive thought patterns. Once the mind expands its perspective, it is impossible to reverse the old thinking.

It's crucial to choose an appropriate quote. Its meaning should express succinctly the truth of the desired belief. Excerpts from great thinkers can be found online and available for everyone. The practice consists of keeping both ideas in mind simultaneously with a steady focus—which supplies the energy for the synaptic reprogramming. A walk or other monotone solitary activity is conducive to cognitive juggling, which cements a commitment to the new meaning. If done diligently, an hour is sufficient to switch to the new belief. However, the change is only temporary; the regular practice should continue until the idea solidifies in the mind.

Learning a demanding musical piece, a foreign language lesson, or doing arts and crafts projects inspires confidence and can lead to positive mental transformation. Therefore, equanimity is the natural and unintended side effect of technical mastery.

Prayer

Religious rituals and prayer have been a part of human existence for millennia. Although the self constantly changes, belief in a Higher Power can fulfill an innate need for permanence and meaning. Today, religion is often considered outdated because a personal appeal to a Higher Power in prayer can reinforce a weakness and abdicate responsibility. Nevertheless, the contemplative opening up to God, the universe, and each other incorporates us into the flow of creation, giving us confidence in its wisdom.

Consultation with an advisor, such as a priest, imam, or rabbi, can provide a higher perspective and is instrumental in motivation and transcendence. Prayer meditation can form steadfast beliefs in supporting goals and the betterment of the self. It builds inner strength by encouraging meaningful, honest efforts. These psychological changes are, perhaps, the most critical, elevating outcomes of prayer. Trust in a higher power creates a mindset unafraid to fail.

Regular meditation or prayer increases satisfaction and joy.

Being Grateful

"A grateful heart is a beginning of greatness. It is an expression of humility. It is a foundation for the development of such virtues as prayer, faith, courage, contentment, happiness, love, and well-being."
— James E. Faust

Social Connections: The strong association of gratitude with happiness gave birth to positive psychology. Belonging and feeling connected are fundamental human needs (Baumeister, 1995). Gratitude is the currency of appreciation granted only to optimistic and trustworthy people. The gesture costs nothing, yet its inspiring, encouraging message improves the social environment.[76] For example, partners who took time to express gratitude in a relationship felt more positive toward the other person (Chang et al., 2022) and felt more comfortable expressing concerns about their relationship. While expressing gratitude can enhance relationships, authenticity is crucial (Algoe et al., 2016). People can discern genuine appreciation from mere flattery.

How we experience the good, the bad, and the ugly dictates our thoughts and behavior. Although stress is an unavoidable aspect of life, we can choose where we focus our attention. Therefore, gratitude is a choice, but it requires humbleness, the affirmation to eliminate pride. Perfection is not the goal; progress is. Like dormant seeds carrying the promise of harvest, subtle positive mental shifts will flower into visible achievements.

Gratitude Training: Those who engage in gratitude training or exercises report less distress, increased satisfaction, and a more profound sense of purpose in life. By bolstering self-confidence and emotional security, it reduces stress. For example, depressed patients expressing appreciation and meaning experienced significant long-term improvement.

Positive Reframing: Studies have demonstrated the benefits of positive reframing. It fosters positive emotions such as joy, enthusiasm, love, and generosity, and improves physical health, including immune function. A strong correlation exists between the absence of negativity and good mental health, such as sleep quality and overall well-being. Shedding emotional baggage allows one to adopt a lighter disposition that improves social aptitude.[77] Compassion allows one to acknowledge the difficulties of others, and altruism compels one to help.

---

[76] Lenz's Law in psychology shows that building a positive environment acts back inspiring in positive motivations in us.

[77] Social aptitude is the ability to relate to others on an emotional level. Children and adolescents with low social aptitude scores are at increased risk of mental health problems.

Interconnectedness of Life: Gratitude underscores the fundamental interdependence and interconnectedness of life. The world unfailingly mirrors our degree of decency.

Gratitude helps set aside egoistic tendencies, paving the way for deeper, more meaningful relationships, improving mental and physical health, and promoting longevity.

Ambition: The Art of Leadership

"A leader will find it difficult to articulate a coherent vision unless it expresses his core values, his basic identity ... one must first embark on the formidable journey of self-discovery in order to create a vision with authentic soul." — Mihaly Csikszentmihalyi

Ambition involves selecting challenges and setting higher goals. It's about having a long-term vision and the audacity to pursue it. The journey of achieving goals fosters self-confidence and growth. Like gradually moving up the mountain, every solved problem improves our confidence in our abilities. Because our attention follows our focus, positive expectations encourage better performance, learning, and creativity. Intrinsic motivation and ambition are essential in personality development and wellness throughout one's lifespan.

Ambition and Leadership: These two qualities are interlinked. While purpose fuels the desire, leadership provides the framework for achieving those desires. It is essential not just for a chief executive officer (CEO); from corporate positions to personal relationships, leadership can unlock individual talent, social dynamism, and enthusiasm.

Team Building

Self-centered leaders trigger defensiveness and discord, but collective ownership depends on collaboration. Exceptional leaders stimulate achievement through motivation rather than manipulation. They respect their colleagues and subordinates. They view learning opportunities as honest mistakes, inspiring loyalty, confidence, and openness. Associates who feel respected can walk through fire for their leaders.

In any business, a team is only as good as its people and as strong as its weakest link. Therefore, helping and mentoring others formulate shared responsibility, motivating productive participation.

Visualization: The most motivated and confident competitor generally wins the Olympic gold medal. A winning mindset does not materialize from thin air. It grows by living in the future of the problem. Immersing in a goal allows one to anticipate challenges and plan solutions.

Female Leadership

Women like Mary Curie[78] and Emmy Noether[79] rose to the top of their professions in male-dominated fields. Nevertheless, until recently, many professional fields remained closed to women. Policy measures to balance the scales, such as gender quotas and blind auditions,[80] have produced varied results. Nevertheless, these initiatives have brought the problem into the public consciousness.

One result of our traditional outlook is that men generally command more respect from both sexes. Some reasons for this are biological. Men are taller and typically have lower-frequency voices, which represents greater power. Although the female voice has deepened over the past decades, women's increasing social, corporate, and political influence reflects more fundamental factors. For example, emotional intelligence (a more feminine trait) inspires openness, trust, and consensus-building in the corporate environment. A diverse leadership representing every color of the human experience is better equipped to navigate the accelerating social and technological progress.

Ultimately, leadership is about believing in one's capability to drive progress and inspire others to join in that journey.

Creativity

"The principal mark of genius is not perfection but originality, the opening of new frontiers." — Arthur Koestler

Creativity is the fuel of achievement. Here, we follow the theoretical basis of this essential skill in Chapter Four. Creativity always appears to arise as a random event that cannot be planned. However, we can take practical steps to improve and cultivate this essential skill.

As earlier sections have shown, mental house cleaning is an emotionally demanding and meticulous task that very often immediately frees mental space for creative ideas. Do not get discouraged if you find unworkable, faulty ideas; if you are persistent and optimistic, innovative solutions will visit you.

---

[78] Mary Curie was a French-Polish chemist with pioneering research on radioactivity.
[79] Emmy Noether was a German mathematician who discovered Noether's theorem.
[80] In blind auditions, the identity of candidates is concealed from the jury by a screen.

Six Steps to Nurture Creativity

1. Understanding the Problem

Problems can present confusing details or contradictions that make understanding appear hopeless. "If you can't explain something in simple terms, you don't understand it," said Richard Feynman. Without identifying the crux of the problem, a solution can remain impossible. Sometimes, phrasing a problem in a question form can uncover a hidden connection, making a solution possible. *Thinking outside the box*, using an interdisciplinary approach, is often more productive than detailed knowledge. A cross-pollination of disciplines has been the source of remarkable accomplishments in art, music, science, and engineering.

2. Finding the Starting Point

Creativity is not a linear process with a convenient recipe that one can follow. Complex problems can be overwhelming. Sometimes, dissecting them into manageable parts offers a convenient starting point toward a solution. Occasionally, better insight is found by stepping back to gain a broader perspective. Other times, it is necessary to dig deep into the roots of the challenge, taking systematic steps.

3. Being Immersed in the Problem

An unwavering commitment to the problem frees one from petty conflicts, allowing an essential insight. Creativity flourishes even in challenging and stressful situations. For example, financial and personal difficulties did not quench the imagination of Mozart and Einstein. Similarly, mental genius is not exclusive to Einstein or Mozart but everyone with a genuine ambition and enthusiastic desire toward a goal.

4. Keeping a Light Disposition

Children are uninhibited in their creative drive, but the freedom of endless playful investigations also benefits adults. Playing has been shown to release endorphins, improve brain functionality, and stimulate creativity (Trezza et al., 2010). Eliminating social pressure and criticism creates a relaxed environment where playful creativity and imagination flourish.

5. Collecting the Unedited Thoughts

Once the ideas flow, their originality must be preserved. Ideas result from the brain wave-like quantum operation, making them fleeting, which may never recur. In addition, short-term memory has enormous energy needs that can tie down the imagination. A journal at the bedside, in the car, or in the purse permits one to write down or sketch ideas as they occur. Even small thought snippets can accumulate into an interconnected whole with originality and direction. We will discuss note-taking in the next section.

6. *Following Up on the Idea*

The non-linear fluctuations of mental time alternate between stressful pressure and expanding insight. In addition, insights are only as good as the diligent, persistent work that follows them. Ideas demand action for fruition. The time to work out the initial inspiration into detailed and wholesome creations can be substantial.

Because creativity is not a straightforward process, success requires perseverance in refining and realizing the concept. Therefore, success often come not the most brilliant creators, but those who never give up. They prevail by learning from mistakes and failures.

Take Notes

> *"Note-taking is important to me: a week's worth of reading notes (or 'thoughts I had in the shower' notes) is cumulatively more interesting than anything I might be able to come up with on a single given day."*
> — Teju Cole

Working memory is fundamental to everyday function as the foundation of complex behavior, reasoning, and planning. The high energy needs of working memory limit its information capacity and temporal range. There are numerous advantages to note-taking.

Memory Augmentation: Note-taking effectively acts as an external extension of our working memory. By jotting down important points, we can ensure that we can revisit and refresh information anytime.

Enhanced Comprehension: Actively taking notes requires processing and understanding the information, which can uncover gaps in our knowledge and deepen comprehension.

Focus: Writing can make one more present during lectures or meetings, potentiating focus and reducing the chances of getting distracted. Summarizing ideas into a concise record guarantees better retention of the information and organizes mental clutter into a crystallized, complete story.

The real fun is in reading the notes later. Periodically reviewing our writing helps to connect related ideas. Gradually improving the outline of the material will generate organization and mental brilliance. There are several popular methods used today.

1. *The Cornell Method*: It allows the visual organization of the material. This structured method encourages active recall and review, helping understand and retain.

2. *The Outlining Method*: It is hierarchical, making it easy to discern main ideas from sub-points. This method can be especially beneficial for linear thinkers or those who prefer a step-by-step breakdown of information.

3. *The Mapping Method*: Visual learners might find this method particularly appealing. By seeing how ideas interconnect, one can get a clearer picture of the subject matter.

Incorporating Technology: While traditional pencil and paper have their charm and effectiveness, the digital age offers a plethora of tools that can enhance the note-taking experience:

Digital Note-taking Tools: Apps like Notion, Roam Research, and Obsidian have become increasingly popular due to their ability to link different ideas and notes, creating interconnected information.

Voice Notes: For those who might find it cumbersome to write or type, voice notes can be a handy tool. Speaking out loud can also reinforce memory.

Digital Stylus and Tablets: Tools like the Apple Pencil with the iPad or the Surface Pen with Microsoft's Surface devices combine the tactile feeling of writing with the advantages of digital storage and organization.

Like taking photographs on your vacation, note-taking helps you remember. Therefore, note-taking isn't just a passive act; it's a dynamic process that aids in understanding, memory retention, and knowledge application. Developing an effective note-taking habit can prove invaluable.

The Need for Renewal

"Predictability can lead to failure." — T. Boone Pickens

After the barren winter, spring brings new life and growth. The changing seasons are perhaps the most poetic representation of renewal. This cyclical process is a constant reminder of the impermanence of phases and the inevitable renewal.

Personal Renewal: It's about bouncing back from setbacks and overcoming challenges. Change is the currency of the future. Adapting to change and learning from experiences are hallmarks of personal renewal.

Renewal can also manifest in personal well-being. Environmental changes, such as moving to a new house, city, or country, inspire fresh perspectives. Nevertheless, activities like meditation, physical exercise, or even a change in diet can rejuvenate the mind and body. For example, self-improvement books, videos, audio, or seminars almost always spur some degree of mental progress. If utilized with sincerity, any new interest, habit, psychological intervention, or practice provides a mental slowing down, resulting in openness and renewal.

Societal and Economic Renewal

Innovation: In the business world, renewal often comes in the form of innovation. Large firms usually protect their successful product line by suppressing upcoming companies. However, the companies that stand the test of time are the ones that continually evolve, innovate and adapt.

Reformation: In societal and governmental contexts, renewal often takes the form of reforms. The power of lobbyists, political donors, and party polarization is inversely proportional to social progress. Societies and nations must constantly reassess and reform their systems and structures to ensure justice, equity, and progress.

The Role of Failure

Learning Opportunities: Every setback is a setup for a comeback. Failures offer unparalleled learning experiences. The most successful individuals and companies have often faced significant disappointments before achieving success.

Pivoting: Failure, which teaches respect for the problem's difficulty, is the best argument to start over. From Plato to Martin Luther King Jr., history's greatest thinkers appreciated the importance of forming a new beginning.

In essence, renewal isn't just about starting fresh but also about embracing the journey of growth and evolution. Whether in personal life, business, or society, periods of renewal are necessary for progress and prosperity. Embracing this concept in thought and action can lead to unparalleled growth and development.

The Menace of Multitasking

"Multitasking the art of doing twice as much as you should half as well as you could. — Anonymous

Our modern existence depends on constant engagement with technology and information flow. Our jobs, schools, and communities encourage us to take on more than we can comfortably handle. Busyness became a modern-day status symbol, an area where we can outdo others and prove our importance. Multitasking offers a seemingly brilliant solution to accomplish several objectives simultaneously. However, our brains aren't wired for simultaneous attention on multiple tasks.

Pitfalls of Multitasking

Reduced Productivity: While multitasking gives an illusion of efficiency, it reduces productivity. Shifting focus between tasks can result in a 40% reduction in productivity, according to some research (Rubinstein et al., 2001).

Memory Impairment: Multitasking can hamper transferring information from short-term to long-term memory, making it harder to retain and recall information.

Increased Stress: Dividing attention between several tasks hinders problem-solving and increases difficulty tuning out distractions (Janet et al., 2020). At the same time, we often miss meaningful signals from our immediate environment. Therefore, while registering distractions, we might miss important information. This can result in heightened stress levels as the brain releases more cortisol, a stress hormone.

Loss of Quality: When attention is divided, the quality of work often suffers. Full attention can only flow toward one subject at any time. Errors are more likely to occur when multitasking than concentrating on a single task. Shoving more things into the same time slot impairs concentration, particularly for long-term multitaskers.

In a world that's always buzzing, there's a counter-cultural but essential need to sometimes step back and single-task. By doing less, paradoxically, we often achieve more. The quality of our work improves, and our mental well-being benefits from this focused approach to tasks.

Benefits of Focused Attention

Deep Work: We lose concentration, and performance declines when working on any task for a long time. As coined by Cal Newport, "deep work" refers to periods of intense concentration without distractions. These periods allow for higher-quality outputs in less time.

Enhanced Creativity: Focused attention allows the brain to think deeply and make connections that might not be evident in a scattered mental state.

Improved Mental Well-being: Mindfulness, the act of being fully present in the moment, is tied to focused attention. Slowing down, switching back and forth between errands, and taking brief breaks to deactivate and reactivate focus can refresh awareness. Mindfulness can reduce stress, increase happiness, and improve overall mental health.

Efficient Learning: Concentrating on one topic at a time allows for more efficient learning and better comprehension.

Fostering Focused Attention: Allocating specific blocks of time to individual tasks. This ensures dedicated attention without the temptation to juggle. Meditation practices can help improve focus and reduce the impulse to constantly switch tasks. Prioritizing tasks based on importance and deadlines allows tackling the most critical chores during peak focus and mental energy periods.

Using Technology

> *"Stimulation of brain pleasure centers can eliminate feelings of rage, fear, and depression."* — James W. Prescott

The advancement of technologies, at the intersection between technology and cognitive, psychological, and behavioral growth, holds tremendous promise for enhancing human capacities and treating disorders. Two areas are up-and-coming for personal development and psychological improvement.

Brain Stimulation in Healthy Subjects: Brain stimulation can safely and effectively enhance mood and boost normal brain function in healthy subjects by changing or even inducing synapses. The process can increase creativity and endurance.

Slow-wave sleep (SWS) is essential in memory consolidation. Electric stimulation during slow-wave sleep potentiates synapses and improves cognitive performance and memory. Brain modulation technology, such as home-administered and remotely supervised protocol, can artificially create or cancel memories and open new, noninvasive avenues to treat mental disorders, depression, PTSD, and others. Similar brain augmentation methods can boost learning and achieve telepathy (brain-to-brain communication). A better theoretical understanding of brain stimulation's long-term effectiveness and side effects will ensure its broader utility.

Using AI for Life Coaching: AI can potentially revolutionize how we approach personal development. With vast data and the ability to analyze individual preferences and needs, AI can tailor self-improvement for physical fitness, cognitive training, and social functioning. The AI life coaching revolution can reach more people, particularly those who could not afford life coaching services due to cost or location, and they are available 24/7. For example, AI life coaches can help identify patterns and triggers contributing to mental health issues and provide personalized strategies and support for individuals to overcome challenges and improve mental well-being.

Ethical Considerations: Nevertheless, there are important ethical considerations. Brain-computer interfaces and AI life coaching could pose significant privacy risks due to their intimate access to personal data. There are also questions on the nature of technology to change or improve human performance. How much should we modify or enhance our abilities? Where do we draw the line between therapy and enhancement? How will access to these technologies influence social inequalities?

Future Directions

Tailored Education: Brain-computer interfaces[81] (BCIs) and virtual reality[82] (VR) could revolutionize education, offering highly personalized learning experiences based on individual neurocognitive profiles.

Emotion Regulation: Future interventions might allow users to regulate their emotional states better, combating issues like anxiety or depression more effectively.

Cognitive Augmentation: As we understand the brain's workings better, there may be opportunities to augment our mental capacities, making us better thinkers, learners, and creators.

Enhanced Interpersonal Communication: In the future, BCIs may enable more profound forms of communication where thoughts and feelings can be shared directly.

---

[81] A brain-computer interface or brain-machine interface is a direct communication pathway between the brain's electrical activity and an external device, which allows people to control machines using their thoughts. These interfaces can help people with disabilities as well as enhance human-computer interactions.

[82] Virtual reality uses an audio-visual headset in 3D virtual reality.

Challenges

Safety: Any invasive procedure, like brain implants, comes with risks. As these technologies develop, ensuring they're safe will be paramount.

Misuse: There's potential for abuse, especially in areas like memory manipulation. How do we ensure these technologies are used ethically?

Dependency: Could there be an over-reliance on AI for life coaching or BCIs for cognitive tasks? What happens if these technologies fail or are unavailable?

The fusion of technology and neuroscience offers a tantalizing glimpse into a future where human capabilities might be significantly enhanced. With the evolution of these powerful tools, society will need to balance the potential benefits with the associated risks, ensuring that these advancements serve humanity's interests.

Shoot from the Hip!

"Action is the real measure of intelligence." — Napoleon Hill

Absolutely! It's essential to acknowledge the danger of venturing into the unknown. The fear of failure and even the risk of success can hold us back. However, embracing challenges and facing them head-on paves the way for growth and innovation.

Negativity and self-doubt can cripple our ability to progress. In the face of adversity, it's easy to craft excuses for holding back, which leads to a cycle of avoidance and stagnation. This mindset leads to seeking short-term comforts and distractions, which might feel good but do not bring lasting fulfillment or growth.

Keeping a Balance: Although people are different in temperament, psychological make-up, and life approaches, we are universal in our vulnerabilities and hopes. Blunders can supply a more advanced understanding, nurturing confidence and courage. True success is not just about crossing the finish line but about the grit, determination, and spirit we show. It's about lighting and keeping that inner fire burning, no matter the odds. Confidence to take charge and seize the moment comes from honest, thorough preparation. So, when challenges arise, we can face them with courage. We must be able to fire.

The Power of Love

"We are shaped and fashioned by what we love." — Goethe

Trust and love are universal themes that have permeated human culture, thought, and behavior since immemorial. *Trusting others and the future leads to generosity, honesty, and cooperation.* Trust is the currency of genuine human interaction. Although trust opens us to potential pain and betrayal, it can lead to deeper bonds, understanding, and cooperation. This element of vulnerability also offers a sense of freedom to live fully. *Without love, the intellect cannot flourish.*

Love's manifestations range from the nurturing love between a parent and child to the romantic love between partners and even the love one feels for one's ideals, God, or a higher purpose. Indeed, religion often serves as a guidepost for our moral compass. Although it requires no money or position, faith is a powerful social capital. It allows tenacity and patience when success (as it often happens in real life) is delayed.

Forming Love: Love originates in trust and self-respect, the source of generosity. This expanding nature engenders a shared attitude, the unity of two exposed, vulnerable hearts, defying the Pauli principle. St Augustine's words "love and do as you will" capture its elevating power. Love in literature, from folktales to contemporary cinema, memorializes its ability to transform and inspire a moral upper ground. Still, every lover discovers love anew — making it as individualistic as snowflakes.

Mating creates oneness, inspiring the caring for the young in emotional animals. In its higher form, romantic love enables the formation of families, where love can grow during our thoroughly dependent earliest years. The failure to develop this trusting love in childhood causes a devastating estrangement from the self. However, the adult heart can develop devotion through a steadfast commitment to love. The fidelity to a person, God, or a high ideal fortifies our trust and replaces weakness and insecurity with love.

Influencing Others: Love and trust also play a pivotal role in influencing others. Leaders like Nelson Mandela and Mahatma Gandhi exemplified how love and trust could break barriers, unite divided communities, and inspire transformative change. Their actions were steeped in love and trust, not just in their followers but also in their ideals and the inherent goodness of humanity.

On the other hand, intimidation and fear might achieve immediate compliance, but they lack the endurance and depth of trust. Authentic leadership and genuine influence stem from a place of love, understanding, and trust.

Lastly, as societies evolve and traditional family structures change, genuine human connection's importance becomes ever more critical. Investing time and effort into nurturing relationships and fostering a positive social environment is indispensable for holistic well-being.

To truly live and find meaning, one must heed the heart's calling, for without love and trust, existence is devoid of depth and purpose.

Conclusions

"Never doubt that a small group of thoughtful, committed citizens can change the world; indeed, it is the only thing that ever has."

— Margaret Mead

Humanity fought and won in the struggle for dominance against the largest, most dangerous animals. Our unique role gives us exclusive responsibilities. We have control over every other species and the whole ecosystem. We have made animals disappear and now may bring them back from extinction. Not the fight, but victory teaches us humility and appreciation. The poor's welfare and the whole biosphere's fate are interconnected.

In India, colonial rule was not defeated by military might but by public opinion. Newspaper articles, which detailed the brutalities of the British army, made colonial rule exceedingly unappealing. The disparities in education and health outcomes resulting from inequality are morally untenable. Contribution to technological and social development requires openness and honesty.

Positive psychological states orient us toward the future, growth, creativity, and joy. These feelings are in line with the expanding universe. Negative sentiments like anger are painful because they contradict the universe's momentum. The mental map on these pages lets you replace insecurity with confidence and wisdom. Serving your mind with all your heart and energy gives you a home turf advantage. Your future is in your hands; if you know how the physical world works, you know how the mind works. Applying just some of the information in this book will put you ahead of the curve.

Life is like a trip. If you do not enjoy the scenery, you waste your time.

As individuals, our bodies have distinct requirements and abilities. We can reach our full potential by honoring and supporting these needs and talents. We should treat ourselves as our most valuable possessions.

Epilogue

"To be yourself in a world that is constantly trying to make you something else is the greatest accomplishment."

— Ralph Waldo Emerson

Religion, the traditional foundation of morality, defines right and wrong by the rules set by God. Evidence indicates the beneficial effects of religion in nearly every aspect of social concern and policy. Yet, the tradition, if not the belief in a higher power, is losing social importance. The satisfied body does not seem to think about the soul. Religion, the moral center of human existence for tens of thousands of years, is becoming secondary to science, the supreme authority over knowledge. Yet, science cannot provide moral guidance, leaving people scrambling to find their motivation. The results have been an epidemic of social alienation and mental diseases.

The ideas presented herein allow you to observe life (including your own) with greater insight and wisdom. The most surprising message is the vast interconnectedness of existence. Matter, mind, and the universe form a unified system where everything touches everything else. Depending on your personal beliefs, you reflect a Higher Power or the Universe, the cosmos' creative intellect. Wisdom is recognizing, understanding, and respecting our mutual inspirations shared by the faithful, atheists, and agnostics.

Elementary school provides primary skills, such as arithmetic and reading, but does not prepare students for social functioning. As corrupted global positioning system (GPS) misleads direction, inappropriate motivation leads to failures and mental disease. The prevalence of social ills indicates a gap in our understanding of emotions and social skills. In achievement, there is no free lunch. The mind projects its thoughts, experiences, and actions forward to dictate what we can and cannot accomplish. Understanding emotional regulation provides confidence and perseverance for achievement.

Progress and innovation require a fluidity of intellect. Emotions influence every aspect of life every day. They affect how a police officer responds to crime, how companies find balance in serving shareholders and customers, and how teachers interact with students and politicians with their constituents.

Technology, arts, and music advance by the brightest and most advanced minds, but social explosions stem from desperation and disadvantage. In a fractured, increasingly unequal world, we cannot isolate ourselves from terrorism and environmental dangers. Global communication brings local wars, religious persecution, and economic hopelessness into the collective consciousness.

The ease of travel and interconnectedness of commerce, social, and cultural institutions increase the possibility of global epidemics, refugee crises, and terrorism. The refugee influx from distant lands contributes to social tension. Temporary fixes cannot possibly stem the expansive and desperate tide. We must address the origin of the oppressive and stagnant conditions in the developing world.

In the traditional view, success transpires at the detriment of others. The achievement and prosperity advocated in this book benefit the whole community. We are the children of the universe. When we embrace our common origin, wealth creation becomes part of collective responsibility. Exceptional achievements require honesty, trust, and a high moral code. These character traits are also preferable for the common good. Thus, individual and social interests center on the same premise.

How do we create a genuinely democratic society? Providing people confidence in their social worth cultivates decency and socially productive behavior. A humane world allows everyone to contribute productively to the common good. I believe that every person has an innate need for belonging, including the need to contribute.

The security against disasters caused by disease and economic misfortunes elevates people and transforms society. Mental diseases and addictions are often the long-term consequences of insecurity. A cooperative and humane society requires basic human security. Because successful people lift the world, abundance, and success should be a human right and a destiny.

Instead of prescribing rigid rules, this book gives you understanding and a higher vision. Its inspiration urges you to better your work and private life. Today, the internet influences every aspect of entertainment, business, learning, and relationships. The internet's democratized access to information can inspire better decisions, from consumer choices to social media and elections. Access to different perspectives allows us to learn new skills and develop social grace.

Before long, new, renewable energy sources might offer inexpensive or free energy. Automatization can expand production based on needs, providing people affordable housing, transportation, and goods. In a decentralized economy, individuals profit directly, exerting personal control and freedom. To take advantage of our new, expanded opportunities requires a high moral code of behavior.

We are part of a coherent cosmic reality in which everything touches everything else. Increasing trust in our social institutions and each other decreases social distance and increases democratic freedom. Comprehending our sophisticated surroundings requires the flexibility to adapt to change. Thus, technological advancement parallels increasing intellectual abstraction and the ability to maneuver highly complex environments.

Our globe is the only boat available to the future, and we all have a valid ticket for the journey. As the Covid-19 pandemic has demonstrated, we cannot save ourselves without saving our global brotherhood. Increasing integration and interdependence of humanity, built on trust, inspire hope, which glues humanity into one giant community motivated by cooperation (Tkadlec et al., 2020), the singleton hypothesis (Bostrom, 2006).

Governments should be concerned for the common good and human flourishing in all areas of life. Because political leaders reflect their electorate, citizens are enormously responsible for this anticipated and necessary climactic transformation. Providing social safety for everyone starts shifting behavior toward social collaboration and generosity.

The highest achievement of our civilization is a web of trust in society.

To The Reader

Thank you for choosing this book for your self-betterment. It laid down a new science with the power to transform our world. Confidence in your abilities provides optimism about the future and makes cheating and ruthless competition pointless. Trust inspires cooperation and generosity, which can become primary motivations for human interaction. In a secure society, protecting the environment becomes natural and second nature.

I introduced a new science of consciousness and emotions. Brain organization and intellect are an organic and inalienable part of the universe. Our continuously shifting life experiences vary, like the surface of the ocean. A sense of expansion feels like the top of the world. In contrast, a sinking, tightening feeling is a sense of insignificance. Maintaining a constant motivation, free of the rises and abates of those emotional tides, is wisdom.

The elegant summary of this book: *only mental progress gives you meaningful, lasting fulfillment*. Positive psychological states help us orient toward the future, mental expansion, creativity, and joy. Negative sentiments, which orient us toward the past, such as anger, are painful. The mental map on these pages about feelings lets you replace insecurity with confidence and wisdom.

As individuals, we are vulnerable to cheating, terrorism, and dictatorial leaders. We must actively work for a better social existence. If you found value in this book, use it daily and pass on the information. You can share an honest review of this book at Amazon, GoodReads, or any other platform. Please join my mailing list to learn about news events, talks, and other activities near you. You will be part of the emerging tide of doers and leaders using these principles.

Disclaimer: This book applies to an ideal society. It does not apply to dictatorial, corrupt, nepotistic regimes.

Afterword

"Nothing great was ever achieved without enthusiasm."

— Ralph Waldo Emerson

My journey began with a failing marriage and severe health issues. I could hardly walk or move at times due to excruciating back pain. Digestive problems and low self-esteem contributed to my anxiety. Although I did not know it, my life spiraled out of control. My husband was also struggling. His bold attempt at forming a biotech company was highly challenging. On occasion, he unleashed his frustration on his family. I felt alone and helpless. My personal life turned into a tunnel, wrapped in darkness, with no way out.

In earlier times, I regularly visited the library with my children. As their independence blossomed, their need for my assistance diminished. The library became my refuge. I sought solace in books about the self, relationships, and meaning. Learning about human behavior alleviated the fog I was living under. I began to observe my husband's struggles in a broader context. I recognized the need to hold onto others is analog to the hold of gravity. Investigating this possible connection spurred my studies in theoretical physics, neuroscience, psychology, and evolution. Those studies confirmed my intuition of the similarities between the organization of the physical world and the mind. My science background helped to formalize my research into a novel hypothesis.

Six years of solitary work resulted in my 2015 book, *The Science of Consciousness*. The work introduces a consciousness hypothesis centering on emotions. It establishes the physical basis of intellect, promising to turn psychology and social sciences more predictable. Considering the mind as a physical system reveals the energy nature of emotions. The fermionic mind hypothesis serves as the basis of this volume. Do not let the name confound you. This is twenty-first-century science.

Future Prospects

"The difficulty lies, not in the new ideas, but in escaping from the old ones." — John Maynard Keynes

I grew up in a religious household where God was supposed to know and observe everything. As a child, I wondered how God could know everything in the past, present, and future. This contrasted with free will because I could not decide to change my mind without God knowing it. The argument presented herein shows that our options relate to how stimulus distorts the brain's energy balance. The meaning of the stimulus determines the size of the energy change of interaction. However, we can choose the direction of energy, whether it triggers endothermic or exothermic processes.

The powerful has a vested interest in keeping the status quo. Nevertheless, the effort to keep back the tides of history is bound to fail. The incoming AI revolution will provide tools to everybody to have a personal assistant, housemaid, or repair technician. However, suppose our free time only increases the hours spent on television and other empty entertainment. In that case, these changes will only generate more suffering. A mental revolution is necessary and inevitable. Greater intellectual awareness will lead to demands for democratization and social transformation.

Assuming that we all share the same space and are 'one', universal spiritual connectedness inspires people to contribute to society according to their abilities.

I firmly believe that we will manage the necessary changes peacefully.

Plato lived in a time when memorization gave great orators the skill to mesmerize and convince their audience. Plato lamented that the development of writing might make remembering unnecessary, a detrimental transformation degrading human intellectual abilities.

Today, we grapple with a similar transformation. The ease of looking up information on anything at any time makes it increasingly unnecessary to remember facts, the purview of scholars and experts. Today, increasingly, imagination, not factual knowledge matters. This might usher in a new era where originality will be valued more than knowledge. According to Einstein's prescient words, "Imagination is more important than knowledge." Perhaps, finding relevant information in seconds will liberate human intellect, allowing everyone to plan, discover, and create.

AI can become the great social equalizer, because anyone with internet access can plan, discover, and create. These abilities, in the wrong hands, can become terrifyingly dangerous. For that reason, learning about emotional regulation is more important than ever. I believe that people can change, and we can make it happen with the right tools. This book contains the tools for such fundamental global transformation.

Darwin's idea of evolution was so shocking that he waited twenty years to present it to the broad community. Although Darwinian evolution became a fundamental principle of biology, it also raised questions about evolution's tendency toward complexity. This book shows the common operational principles of physical and biological systems, evolution, psychology, and social sciences. We formulate emotional separation from our loved ones or neighbors based on physical distance. The fermionic mind hypothesis posits unexpected connections supported by well-respected studies.

The FMH is the first hypothesis offering a physical basis for motivation and human behavior. It is also the first overarching idea encompassing physics, the mind, and evolution. The universe is made up of orthogonal spatial and temporal energies. The existence of the particle organization on three different levels of the universe (the material fermion, the emotional fermion, and the universe) explains power-law distributions in geology, physics, population dynamics, biology, sociology, and other fields.

FMH supports humanity's intuition throughout the ages and cultures about an underlying unity of all existence. In agreement with the Aristotelian physical worldview, physical reality abhors a vacuum. Matter and the universe have symmetric energy structures. Still, the mind is positioned between matter and the universe, with characteristics of both. The negative attitude mind is entirely directed by the environment (making it matter-like). Only the positive attitude mind has free will (making it God-like). However, the positive attitude has no incentive or motivation to change. The idea that evolution, including social development, is guided toward complexity and order suggests a shining future for the cosmos' intelligent occupants, including mankind.

Some of the expressions you find in this book:

Anxiety: Anxiety is the emotional state of not having options.

Consciousness: Consciousness is the awareness of being a separate entity within the environment.

Emotions: Emotions (sentiment or feeling) are the brain's energy states that trigger appropriate behavior to recover the energy-neutral resting state.

Emotional Gravity: Emotional gravity (temporal gravity) is the strength of attachments toward people and things.

Emotional or Social Temperature: Emotional temperature is the social equivalent of temperature in physics. As particle collisions form temperature, brain frequencies determine free will.

Emotional Distance: Emotional distance is the degree of trust. Sizeable emotional separation spares one from attitude effects (emotional magnetism, such as the Pauli exclusion principle). We form conceptual distance to abstract entities, whereas emotional distance characterizes relationships.

Festinger's Cognitive Dissonance Theory: Conflicting attitudes, beliefs, or behavior produces a feeling of mental discomfort, leading to an alteration in one's opinions or actions to restore balance.

Interaction: Mental interaction is energy/information exchange with the environment via stimuli. The resulting synaptic changes modify the brain's energy state. In this way, the brain participates in the environment's energy cycle.

Intellect: Intellect is the ability to comprehend the environment with foresight. Abstraction is the ability to maneuver highly complex environments.

Intelligent Systems: Intelligent systems are aware of their past and use them to solve problems and utilize energy efficiently. Engaging with the surroundings in flexible ways increases opportunities in the future.

Mental Energy: Mental energy is an abstract mental quality that permits happiness, confidence, focus, motivation, and productivity. It is a necessary ingredient of achievement.

Mental freedom is the ability to act uninhibited. Freedom correlates with survival needs in biological systems, such as air, food, or water. For example, a good lunch increases liberty, testifying about the time-dependence of life.

Psychological Lensing: Psychological lensing generates two different answers about the same event.

Social Energy: Social energy is a special form of motivation boosting decision-making.

Stress: Temporal pressure resulting from positive temporal curvature.

Temporal: Temporal concerns time.

Temporal Field: The social environment represents the local curvature of the temporal field. Therefore, it determines the perception of time.

Temporal curvature (field strength) corresponds to information-loaded, automatic thinking patterns.

***Temporal Gravity*:** Temporal gravity shows a close correlation with social status.

***The Temporal Landscape*:** The willingness for interaction depends on the energy state of synaptic connections, which reflect the individual's history. Our behavior correlates with our mental state because our memories form a topological surface that influences the neuronal activation path's trajectory.

***Temporal Pressure (Stress)*:** Decreasing the volume of gases increases particle collisions and temperature. In biological systems, the dependence on physiological needs determines our welfare. The shortage of air, water, food, and money increases alertness for survival. The stressful, heightened mental state can trigger desperate, dangerous, risky, or even lawless actions.

***Uncertainty*:** Uncertainty is having several possible alternative explanations for personal histories. Uncertain individuals cannot tolerate uncertainty in the environment.

Cognitive Behavioural Therapy (CBT): In CBT, the so-called ABC Technique of Irrational Beliefs is used. In life, there can be an activating event (A) or situation that triggers a negative thought based on beliefs (B) in a patient, which then evokes an emotional response in the patient. The patient then acts on these thoughts and emotions, which result in a consequence (C).

The therapist will help the patient interpret the event, showing that his/her interpretation is incorrect.

Appendix

This book introduces an innovative theory of consciousness known as the "fermionic mind hypothesis." It presents numerous compelling examples that validate this theory in individual and social behavior and highlights the growing significance of quantum cognition across various fields, including language, decision-making, memory, reasoning, judgment, perception, social sciences, and economics. Scientific literature has demonstrated and confirmed several predictions in my 2015 book, "The Science of Consciousness." For example, the physical foundation of sentiment contagion led to a detailed understanding based on fluid dynamics. Moreover, many more instances of validation can emerge in the future.

In the following, the fermionic mind hypothesis is pitted against the physical characteristics of fermions using category theory from abstract mathematics. Category theory, which studies structures and the relationships between them, offers a panoramic perspective that can demystify connections between diverse disciplines, including neuroscience, psychology, general relativity, quantum theory, and mathematics (Anderson, 2021; Ehresmann and Gómez-Ramírez, 2015; Phillips, 2022). As a result, similarities or equivalence between fermions and consciousness become discernible.

At the core of category theory lays the Yoneda lemma, postulating that an object's nature can be fully apprehended by scrutinizing its relationships and interactions with other entities (Tsuchiya and Saigo, 2021). This lemma enables equivalence between objects A and B, up to isomorphism, contingent on their relational dynamics within a category, implying that they share the same structure. The object's structure serves as a guide to its essence, offering a valuable tool for unraveling complex and elusive phenomena, such as consciousness. In this context, a morphism is an oriented "relationship," "process," or "transformation" between objects.

By applying category theory and the Yoneda lemma to the fermionic mind hypothesis, we establish a mathematical foundation that elucidates the relational interplay between consciousness and material systems (Tsuchiya and Saigo, 2021). This innovative framework can verify the hypotheses by aligning harmoniously with existing empirical evidence.

Temporal Orientation: The Orthogonality of Sensory Reality and Cognition

The nature of consciousness, whether a continuous stream of experiences, emotions, or thoughts occurring at specific moments, has been a longstanding and intriguing question among philosophers, psychologists, and neuroscientists. However, there is a significant duration of unconscious processing before conscious perception and awareness emerge. Moreover, despite our continuous flow of experiences, we remember episodes as discrete sequences of events (Shapiro, 2019) and see the future as steps on the temporal measuring stick of our imagination, which is consistent with discrete processing. Thus, consciousness shows continuous and discrete qualities and a framework that reflects both might represent how we perceive the world around us.

For example, directly manipulating attention postdictively, i.e., at short time delays (Shen et al., 2020), influences choice. Therefore, cognition is not instantaneous, moment-by-moment consciousness constructions but incorporates information after perception by the so-called postdictive effect. In the above example, substantial periods of continuous unconscious processing precede discrete conscious percepts (Herzog et al., 2020). Likewise, despite the constant flow of experience, we remember memories as discrete sequences of events (Shapiro, 2019).

Next, we turn to the cognitive process. Conscious experience is intrinsically tied to the "now," representing an unbroken stream of awareness, seamlessly transitioning from one moment to the next. Nevertheless, each present moment uniquely represents discrete conscious experiences of successive "nows" that become memories. This twofold nature of consciousness signifies a constant shift in focus, where observation changes its underlying parameters, making it impossible to grasp—analogous to the observer effect. In quantum mechanics, the observer effect is the change of an observed system by observation. It also suggests that investigation via measurement causes the discrete collapse of the wave function, the so-called wave-particle duality.

The lens of a "monoid" can effectively examine the conscious duality concept. A monoid can encompass multiple morphisms, each representing the ongoing flow of consciousness. In a monoidal category where only the identity morphism exists, the arrow returns to itself, symbolizing the continuous unfolding or extension of conscious experience (Figure 1).

$$id_a \circ f = f = f \circ id_a$$

The concept of consciousness as a monoid closely aligns with the brain's ability to coalesce disparate and disorderly information into a unified perception (Tring et al., 2023). Unity aligns along a temporal arc (Smallwood et al., 2021) that endows intellect with a remarkable predictive ability (Northoff et al., 2019). "Therefore, time touches the eternal now at each moment,… disappears moment by moment and is born moment by moment… as a continuity of discontinuity (Nishida, 1948, p. 342, after Taguchi and Saigo, 2023)." Similarly, with time, consciousness stays constant yet continuously updates, shaping our social insights, thoughts, and actions (Smitha et al., 2017; Wolff et al., 2019).

Morphisms of consciousness can be viewed from the perspective of the monoid, where the morphism is always from the same object to the same object. "One morphism and another, one morphism and itself, can all be composed because the morphism of a monoid has only one object, which means that all morphisms ('passages') are connected at the same object ('now') because their starting points (domains) and endpoints (codomains) are the same objects (Taguchi and Saigo, 2023)." The flow of consciousness perpetually remains in the same place because it originates from the same object as their domain and returns to it as their codomain. In essence, the dynamic, streaming aspect of the present corresponds to the diverse morphisms within a monoid. In contrast, the static, standing aspect aligns with the unique object of the monoid. This monoid framework effectively captures consciousness' intricate structure, encompassing its ever-changing and constant facets (Figure 2).

Now, we return to the orthogonality question. The brain's spatial data compression formulates an orthogonal sensory temporal hologram (Saaty and Vargas, 2017), representing an orthogonal projection onto the two-dimensional cortical surface (Déli, 2020a,b). In neuroscience, the hippocampus' place cells transform spatial relationships into a temporal projection (Shimazaki, 2020), with the intermediate hippocampal CA1 serving the spatial aspects and the dorsal hippocampal CA1 serving the temporal aspects of episodic memories (Barker et al., 2016), and coordinate collective behavior (Forli and Yartsev, 2023).

Figure 1. Conscious (*a*) as a monoid. The arrow represents the identity morphism.

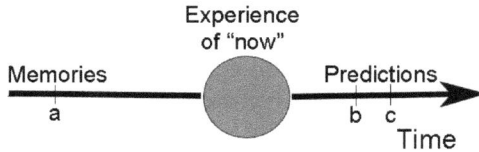

Figure 2. Consciousness as a continuous and discrete temporal flow. Memory and expectation materialize at discrete temporal locations marked: time points a, b, c.

The arrow of time, a fundamental concept in physics, demonstrates a profound connection to the rate of entropy generation (Lucia and Grisolia, 2020). The brain's ability to consistently form endothermic processes entails entropy generation by enhancing the degrees of freedom (Shi et al., 2019; Yang et al., 2019). Therefore, intellect is an orthogonal orientation vis-à-vis material structure, where entropy generation represents complexity and order.

In category theory, orthogonality defines relationships between objects and morphisms within categories. For example, in the category of sets, two sets are considered orthogonal if they have an empty intersection. In a class of morphisms X in a category C, the orthogonal complement of X within C is a class of morphisms Y in C. These morphisms Y in C are such that for each morphism f: X → Y in C × C, there is a unique morphism g: A → B in C such that f is orthogonal to g. Therefore, for a composable pair of morphisms (f,m), there is an operation → called "composition" that makes the morphism m ○ f called the "composite of m and f." Likewise, for the composable pair of morphisms (e,g), the operation ○ called "composition" makes the morphism g ○ e called the "composite of g and e." The domain of m ○ f and g ○ e coincide with the domain of d, and the codomain of m ○ f and e ○ g coincide with the codomain of d.

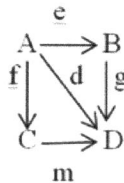

Two morphisms e: A → B and m: C → D are orthogonal if for any f: A → C and g: B → D, there exists a unique morphism d: A → D such that m ∘ f = g ∘ e.

The Mind's Particle Nature and the Resting Equilibrium

So far, we have shown that consciousness is a temporal, complexity-generating system ensuring a rich variety of reference frames (Gaio and Rizzuti, 2022). Category theory can differentiate between different reference frames. Like the gravity field governs particle movement, temporal orientation defines our cognitive freedom. Temporal orientation rests with the resting state. Like particles following the geodesic dictated by gravity, the resting state's mental expanse functions as a steady force shaped by memories and experience (Li et al., 2019).

Another intriguing analogy is the so-called emotional electromagnetism, which rapidly oscillates between attraction and repulsion (Drigotas, 1993). For example, a dog's wagging tail and raised hackles are analogous to electric polarities, an instant feeling for or against it (Surov, 2022). Emotions action-producing capacity (Nazmi et al., 2020) is a spin-like, positive, or adverse energy condition (Kao et al., 2015; Deli, 2023) supported by the hormone oxytocin (Anpilov et al., 2020) (Table I). Partiality can spread to social groups, creating divisions (Krems et al., 2023).

The dilation of time perception in negative experiences potentiates their action-producing power while averting action in positive emotions (Remmers and Zander, 2018; Rudd et al., 2012), underlying their energy nature. Emotions are the forces of motivation.

Quantum Physics and General Relativity Characterizes Human Thought

Psychology often turns to classical physics to explain individual or social behavior. Sentiment Contagion explains the spontaneous spread of emotions and related behaviors in social groups by heat conduction (Goldenberg et al., 2018). Social media groups can act as Filter Bubbles that confirm or enhance individual bias (Holone, 2016). Finally, hysteresis can explain aggravation, unemployment (Plotnikov, 2014), or the mob.

Adherence to resting balance represents particle-like stability, which makes it impossible to govern or predict our emotions and behavior. Like scattering experiments uncovering the atom's internal structure, perceptions expose our ineffable, private, and inaccessible mental attitude structures. The mind's abstract world model flexibly adapts to different situations and experiences (Tring et al., 2023; Wolff et al., 2021). Unruly thought processes (Zanin et al., 2020) converge into a coherent self-narrative (Smitha et al., 2017), the foundation of a subjective, transcendental, and uniquely personal psychological equilibrium (Kolvoort et al., 2020).

Surgical separation of the brain's two hemispheres leads to the famous split-brain experience of distinct personalities (Pinto et al., 2017), but further partitioning of the brain annihilates consciousness, indicating that the mind is fundamental and irreducible: the elemental unit of intellect (Déli, 2020a); see Table I. Nevertheless, the well-defined observational outcomes of cognition emerge from the dynamic intrinsic activation patterns on the cerebral cortex (Pang et al., 2023; Selesnick & Piccinini, 2018). These fluctuations and uncertainties profoundly influence arousal (Anderson et al., 2019), surpassing the impact of external stimuli (Dempsey et al., 2022). The wave-like characteristics of these fluid and vague thoughts, emotions, and concepts settle into discrete convictions and beliefs (Herzog et al., 2020), akin to the complementarity principle in quantum physics (Basieva et al., 2019),

Quantum cognition uses the mathematical principles of quantum theory to explain comprehension and decision-making (Deli, 2020.a,b; Khrennikov, 2020; Tring et al., 2023; Wolff et al., 2021). The indistinguishable intrinsic properties of fermions, such as charge, mass, and spin, find emotional analogies, allowing us to identify with others and form comparisons.

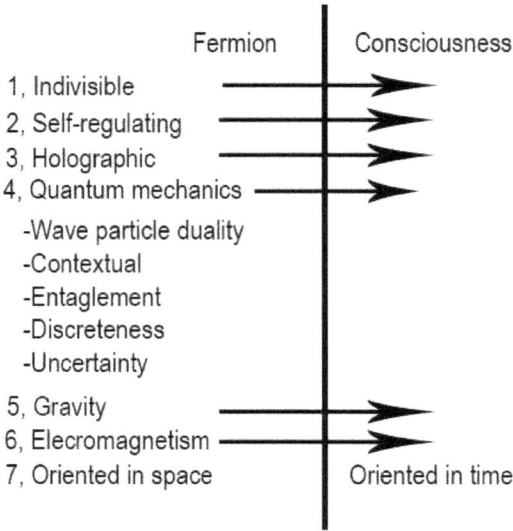

Fermion	Consciousness
1, Indivisible	
2, Self-regulating	
3, Holographic	
4, Quantum mechanics	
-Wave particle duality	
-Contextual	
-Entaglement	
-Discreteness	
-Uncertainty	
5, Gravity	
6, Elecromagnetism	
7, Oriented in space	Oriented in time

Table II. The orthogonality of material fermions and the temporal mind. The mathematical formalism of quantum mechanics describes both systems. Still, their opposite high entropy manifestations mean a disorder in material objects but order and intellect for the mind.

	Matter Fermion	Temporal Fermion (the mind)
Unity	The smallest unit of matter	The smallest unit of intellect
Quantum mechanics	Applies in space	Applies in time
Entropy	The arrow of time	Future orientation
High entropy	Disorder	Order, creativity, intellect
Physical existence	Space	Time
Corresponding field	Gravity	Social manifold
Lorentz transformation	Time dilation	Dilation of time perception
	Gravity	Stress (emotional gravity, negative emotions)
	Acceleration	Mental expansion (awe and other positive focus)
Pauli exclusion principle	Forms material structure	Causes social hierarchy
Thermodynamic outcome	Exothermic cycle	Endothermic cycle
Particle-like stability	Particles are unchanging	A constant sense of self throughout life

Fermions' ability to form stable objects is due to their adherence to the Pauli exclusion principle, which prevents them from simultaneously occupying the same quantum state within an atom, causing their resistance to compression. This is analogous to the human tendency for competition, contradiction, and criticism, the source of territorial needs in the animal world and societal hierarchical structures (Saaty and Vargas, 2017; Wato et al., 2020). Therefore, the quantum qualities become the foundation for classical behavior.

Classical Analogies in the Brain's Operation

The human brain is an intricate system of neural networks, electrical impulses, and complex cognitive processes. Mental balance is highly dynamic, yet it can give rise to the resting equilibrium (Zanin et al., 2019). The memory dependence of the brain's resting state turns it into a temporal field, shaping our thoughts and actions even beneath the threshold of consciousness. The resting state, characterized by coherent low-frequency resting oscillations, provides many degrees of freedom, allowing intellectual capacity (Yang et al., 2019) and creativity (Shi et al., 2019). It encodes individual and social expectations to impact what we perceive, hear, and see (Xu and Schwarz, 2018) and catalyze behaviors between creativity and self-destructive tendencies (Beall and Tracy, 2017). The social world represents emotional gravity.

The energy of attachments dictates our loyalty to people, places, and habits. Losses are profound reminders of attachment power, which makes their true strength painfully evident. The time pressure of stress increases emotional temperature, the social analog of temperature (Stewart and Plotkin, 2012). Its cognitive burden erodes our mental energy reserves and diminishes intellectual capacity (Makharia et al., 2016; Mani et al., 2013). On the other hand, supply abundance lowers social temperature (Tkadlec et al., 2020).

Akin to gravity, emotional gravity regulates the individual's mental landscape. Inversely, people within a network influence the social structure of their surroundings, thereby impacting the entire interconnected web (Li et al., 2019). This interconnectedness highlights the broader social dynamics of the brain operation. Like the relationship between the gravity field and its particles, the social sphere intersects the personal world.

Category theory provides tools to differentiate between frames of reference (Gaio and Rizzuti, 2022). The temporality of consciousness is an orthogonal transformation t between consciousness G and the fermions F (Figure 3), hom (x, _): G ≅ hom (x, _): F (Table II).

Emotional electromagnetism is another intriguing analogy that can be drawn from the brain's operation. Emotions' multidimensional representations and varied cultural and brain activity profiles hide their involuntary nature and action-producing ability (Nazmi et al., 2020). Therefore, like photons, emotions represent energy (Kao et al., 2015); see Figure 4.

Consciousness (C) and fermions (D) have some shared characteristics (Figure 4). A functor F from C to D is a mapping that associates each object in C to an object in D.

$$F: C \rightarrow D$$

C: f̲(QM, Unity, Fundamental forces, Classical laws)
D: f̲(QM, Unity, Fundamental forces, Classical laws)

$$Hom_C\ (\_,A) = Hom_D\ (\_,A)$$

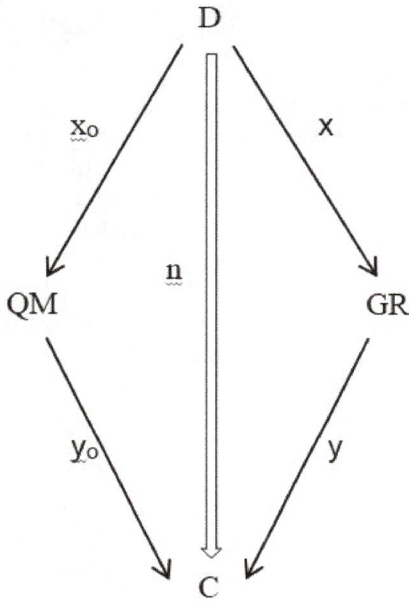

Figure 3. The isomorphism of consciousness and fermions. Quantum mechanics and relativity were developed to describe fermions. Neuroscience in the past decades has shown that consciousness shows both quantum and classical characteristics. Therefore, n can be obtained by the isomorphism of consciousness (C) and fermions (D) by composition.

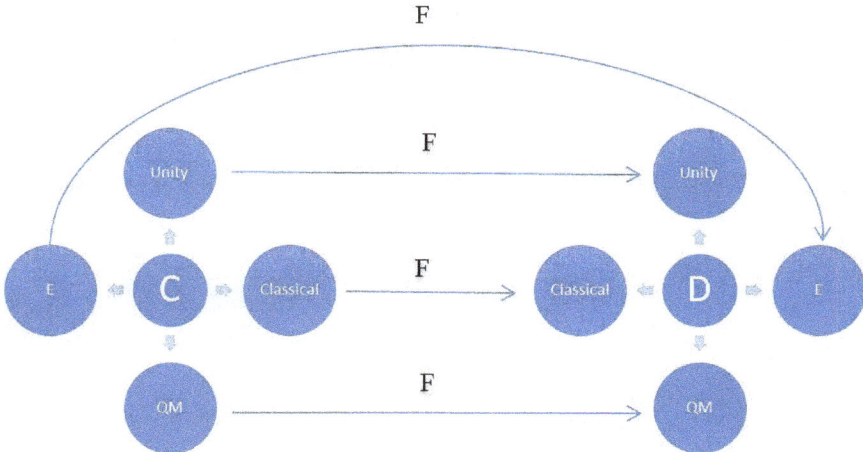

Figure 4. Applying the Yoneda lemma to consciousness. Consciousness (C), fermion (D), quantum mechanics (QM), Emotional forces (E), and Classical behavior (Classical) Consciousness and fermions show almost identical organization and operational principles.

Conclusions

Physics and cognitive science have their concepts of objects and processes, i.e., morphisms, organized into categories with many standard features. Consciousness and artificial general intelligence (AGI) should exhibit a similar abstract structure, forming an isomorphic relationship. The likely analogy between artificial or neural-based intellect is due to their physical organization. A strong argument and mathematical verification support the fermionic mind hypothesis; its fate ultimately depends on further testing against the physical universe.

While our memory library evolves, emotions emerge as the driving forces behind the motivation to maintain the resting state, the anchor for our sense of self. Much like the permanence of particles, the resting state ensures the continuous identity of the self. Furthermore, our memories are a temporal field that guides our thoughts and actions, even when these processes remain beneath the threshold of consciousness. The duality of consciousness bears a striking resemblance to the wave-particle duality observed in quantum mechanics. Moreover, consciousness' temporal nature can be viewed as orthogonal to physical processes, representing an endothermic quality.

Recognizing the mind's isomorphism with fermions provides historical, philosophical, and psychological comfort. We are the children of the universe and our environment, but our words and actions formulate the world. The fermionic mind hypothesis represents a significant step forward in understanding consciousness and its implications. It can significantly aid AI research, psychology, and social sciences.

References

Aberg, K.C., Toren, I. and Paz, R. (2023). Irrelevant threats linger and affect behavior in high anxiety. *The Journal of Neuroscience,* 43: 656–671.

Ahmed, M.S., Priestley, J.B., Castro, A., Stefanini, F., Solis Canales, A.S. and Balough, E.M. (2020). Hippocampal network reorganization underlies the formation of a temporal association memory. *Neuron,* 107(2): 283–291. e6.

Algoe, S.B. and Zhaoyang, R. (2016). Positive psychology in context: effects of expressing gratitude in ongoing relationships depend on perceptions of enactor responsiveness. *J. Posit. Psychol.,* 2016;11(4): 399–415.

Alhussien, M.N. and Dang, A.K. (2020). Interaction between stress hormones and phagocytic cells and its effect on the health status of dairy cows: A review. *Veterinary World,* 13(9): 1837–1848.

Anderson, E.C., Carleton, R.N., Diefenbach M., Han, P.K.J. (2019). The relationship between uncertainty and affect. *Front Psychol.,* 12;10: 2504.

Anderson, C., Hildreth, J.A.D. and Howland, L. (2015). Is the desire for status a fundamental human motive? a review of the empirical literature. *Psychological Bulletin,* 141(3).

Anderson, J.R. (2000). *Learning and Memory: An Integrated Approach.* New York: John Wiley and Sons.

Anpilov, S., Shemesh, Y., Eren, N., Harony-Nicolas, H., Benjamin, A., Dine, J. and Oliveira, V.E. (2020). Wireless optogenetic stimulation of oxytocin neurons in a semi-natural set-up dynamically elevates both prosocial and agonistic behaviors. *Neuron.*

Aristotle, Metaphysics, 350 B.C.E.

Baik, J.H. (2020). Stress and the dopaminergic reward system. *Exp. Mol. Med.,* 52: 1879–1890.

Basieva, I., Cervantes, V.H., Dzhafarov, E. and Khrennikov, A.Y. (2019). True contextuality beats direct influences in human decision making. *Journal of Experimental Psychology.*

Baumeister, R. and Leary M. (1995). The need to belong: desire for interpersonal attachments as a fundamental human motivation. *Psychological Bulletin,* 117(3): 497–529.

Baumeister, R.F. (1984). Choking under pressure: Self-consciousness and paradoxical effects of incentives on skillful performance. *Journal of Personality and Social Psychology,* 46: 610–620.

Beall, A.T. and Tracy, J. (2017). Motivational psychology: how distinct emotions facilitate fundamental motives. *Social and Personality Psychology Compass,* 11.

Bechler, C.J., Tormala, Z.L. and Rucker, D.D. (2019). Perceiving attitude change: How qualitative shifts augment change perception. *Journal of Experimental Social Psychology.*

Belmans, E., De Vuyst, H., Takano, K. and Raes, F. (2023). Reducing the stickiness of negative memory retrieval through positive memory training in adolescents. *Journal of Behavior Therapy and Experimental Psychiatry,* 81: 101881.

Berkman, E.T. (2018). The neuroscience of goals and behavior change. *Consulting Psychology Journal,* 70(1): 28–44.

Bethell, E.J., Holmes, A., MacLarnon, A. and Semple, S. (2012) Evidence that emotion mediates social attention in rhesus macaques. *PLoS ONE*, 7(8): e44387.

Birney, D.P. and Beckmann, J.F. (2022.) Intelligence is cognitive flexibility: why multilevel models of within-individual processes are needed to realise this. *J. Intell.*, (Aug.) 1, 10(3): 49.

Bolló, H., Háger, D.R., Galvan, M. and Orosz, G. (2020). The role of subjective and objective social status in the generation of envy. *Frontiers in Psychology*, 11.

Bonwell, 1991. *Higher Education Report No. 1*, GWU, Washington, DC.

Brighi, E. and Cerella, A. (2015). An alternative vision of politics and violence: introducing mimetic theory in international studies. *Journal of International Political Theory*, 11: 25–33.

Brinch, C.N. and Galloway, T.A. (2012) Schooling in adolescence raises IQ scores. *Proc. Natl. Acad. Sci.*, 109: 425–430.

Britton, W.B., Lindahl, J.R., Cooper, D.J., Canby, N.K. and Palitsky, R. (2021). Defining and measuring meditation-related adverse effects in mindfulness-based programs. *Clin. Psychol. Sci.*, 9(6): 1185–1204.

Brummelman, E., Thomaes, S., Orobio de Castro, B., Overbeek, G. and Bushman, B.J. (2014). The adverse impact of inflated praise on children with low self-esteem. *Psychological Science*, 25(3): 728–735.

Buch, E., Claudino, L.M., Quentin, R., Bönstrup, M. and Cohen, L.G. (2021). Consolidation of human skill linked to waking hippocampo-neocortical replay. *Cell Reports*, 35: 109193–109193.

Carmona-Halty, M., Salanova, M., Llorens, S. and Schaufeli, W. (2019). How psychological capital mediates between study–related positive emotions and academic performance. *Journal of Happiness Studies*, 20: 605–617.

Carr and Schatman. (2019). *American Journal of Public Health*, 50–51.

Chandler, C.C. (1989). Specific retroactive interference in modified recognition tests: evidence for an unknown cause of interference. *Journal of Experimental Psychology: Learning, Memory, and Cognition*, 15(2): 256–265.

Chang, L.W., Gershman, S. and Cikara, M. (2019). Comparing value coding models of context-dependence in social choice. *Journal of Experimental Social Psychology*, 85: 103847.

Chang, Y.P., Way, B.M., Sheeran, P., Kurtz, L.E., Baucom, D.H. and Algoe, S.B. (2022). Implementation intentions to express gratitude increase daily time co-present with an intimate partner, and moderate effects of variation in CD38. *Sci. Rep.*, 12: 11697.

Chen, T., Wang, P., Wang, Y. and Irish, M. (2023). Combining experience sampling with temporal network analysis to understand inertia of negative emotion in dysphoria. *Journal of Affective Disorders*, 338: 246–253.

Conant, R.C. and Ashby, W.R. (1970). Every good regulator of a system must be a model of that system. *International Journal of Systems Science*, 1: 511–519.

Connelly, B.D., Bruger, E.L., McKinley, P.K. and Waters, C.M. (2017). Resource abundance and the critical transition to cooperation. *J. Evol. Biol.*, 30: 750–761. pmid: 28036143.

Cook, R. et al. (2012). Automatic imitation in a strategic context: Players of Rock-Paper-Scissors imitate opponents' gestures. *Proc. R. Soc. B: Biol. Sci.*, 279: 780–786.

Cox, R.T. (1979). Of inference and inquiry, an essay in inductive logic. *In: Proceedings of the Maximum Entropy Formalism—First Maximized Entropy Workshop*, Boston, MA, USA, 119–168.

Crawford, L., Knouse, L., Kent, M., Vavra, D.T., Harding, O., Leserve, D. et al. (2020). Enriched environment exposure accelerates rodent driving skills. *Behavioural Brain Research*, 378.

Crossley, M., Benjamin, P.R., Kemenes, G., Staras, K. and Kemenes, I. (2023). A circuit mechanism linking past and future learning through shifts in perception. *Science Advances*, 9.

Csikszentmihalyi, M. (2080). *Flow: The Psychology of Optimal Experience*. New York, NY: HarperCollins.

Cunnane, S.C., Trushina, E., Morland, C., Prigione, A., Casadesus, G. and Andrews, Z.B. (2020). Brain energy rescue: An emerging therapeutic concept for neurodegenerative disorders of ageing. *Nature Reviews. Drug Discovery*, 19(9): 609–633.

Dabney, W., Kurth-Nelson, Z., Uchida, N., Starkweather, C.K., Hassabis, D., Munos, R. and Botvinick, M.M. (2020) A distributional code for value in dopamine-based reinforcement learning. *Nature*, 577: 671–675.

de Botton, A. (2004). *Status Anxiety*. New York, NY: Pantheon Books.

Deli, E. (2015). The Science of Consciousness self-published.

Deli, E.K. (2023). What Is Psychological Spin? A Thermodynamic Framework for Emotions and Social Behavior. Psych., 5(4): 1224–1240.

Deli, E. and Kisvarday, Z. (2020). The thermodynamic brain and the evolution of intellect: the role of mental energy. *Cognitive Neurodynamics*, 14(6): 743–756.

Déli, E. (2020b). The thermodynamic implications of the FMH. *Activitas Nervosa Superior*, 62(3): 96–103. DOI: 10.1007/s41470-020-00074-0.

Déli, E. (2020a). Can the FMH explain consciousness? the physics of selfhood. *Activitas Nervosa Superior*, 62: 35–47. DOI: 10.1007/s41470-020-00070-4.

Déli, E., Peters, J. and Kisvárday, Z. (2021). The thermodynamics of cognition: a mathematical treatment. *Computational and Structural Biotechnology Journal*, 19(2).

Déli, E., Peters, J. and Tozzi, A. (2018). The thermodynamic analysis of neural computation. *J. Neurosci. Clin. Res.*, 3, 1.

Dempsey, W.P. Du, Z., and Nadtochiy, A. (2022). *PNAS*, 119 (3), e2107661119.

Dennett, D.C. (2018). Facing up to the hard question of consciousness. Philosophical Transactions of the Royal Society of London. Series B. *Biological Sciences*, 373(1755): 20170342.

Di Domenico, S.I. and Ryan, R.M. (2017). The emerging neuroscience of intrinsic motivation: a new frontier in self-determination research. *Front. Hum. Neurosci.*, 11: 145.16.

Diener, E., Suh, E.M., Lucas, R.E. and Smith, H.L. (1999). Subjective well-being: three decades of progress. *Psychological Bulletin*, 125(2): 276–302.

Dijksterhuis, A.J., Bos, M.W., Nordgren, L.F. and van Baaren, R.B. (2006). On making the right choice: the deliberation-without-attention effect. *Science*, 311: 1005–1007.

Doerig, A., Sommers, R.P., Seeliger, K., Richards, B.A., Ismael, J., Lindsay, G.W. et al. (2022). The neuroconnectionist research programme. *Nature Reviews Neuroscience*, 24: 431–450.

Drigotas, S.M. (1993). Similarity revisited: a comparison of similarity-attraction versus dissimilarity-repulsion. *British Journal of Social Psychology*, 32(4): 365–377.

Duhigg, 2012. *The Power of Habit*. Random House.

Duyme, M., Dumaret A-C. and Tomkiewicz, S. (1999). How can we boost IQs of "dull children"? a late adoption study. *Proc. Natl. Acad. Sci.*, 96: 8790–8794.

Einhäuser, W., Stout, J., Koch, C. and Carter, O.L. (2008). Pupil dilation reflects perceptual selection and predicts subsequent stability in perceptual rivalry. *PNAS*, 105(5): 1704–1709.

Elliott, A.K. (1971). *The Sociology of Occupations. Professions and Professionalization*. Edited by J.A. Jackson; and *Status Passage: A Formal Theory*. Barney Glaser and Anselmn L. (2020). Cambridge.

Elphinston, R.A., Sullivan, M.J., Sterling, M., Connor, J.P., Baranoff, J., Tan, D. et al. (2021). Pain medication beliefs mediate the relationship between pain catastrophizing and opioid prescription use in patients with chronic non-cancer pain. *The Journal of Pain*, 23(3): 379–389.

Farmer, C.G. (2020). Parental care, destabilizing selection, and the evolution of tetrapod endothermy. *Physiology*, 35(3): 160–176.

Festinger, L. (1957). *A Theory of Cognitive Dissonance.* Stanford University Press.

Fisher, R.S., Acharya, J.N., Baumer, F.M., French, J.A., Parisi, P. and Solodar, J. (2022). Visually sensitive seizures: an updated review by the Epilepsy Foundation. *Epilepsia,* 63: 739–768.

Forli, A. and Yartsev, M.M. (2023). Hippocampal representation during collective spatial behaviour in bats. *Nature,* 621: 796–803.

Fredericks, C.A., Sturm, V.E., Brown, J.A., Hua, A.Y., Bilgel, M. et al. (2018). Early affective changes and increased connectivity in preclinical AD's disease. *ADs Dement (Amst).,* 10: 471–479.

Fry, R., (2017). Physical intelligence and thermodynamic computing. *Entropy,* 19: 107.

Gao, Z., Cui, X., Wan, W., Zheng, W., Gu, Z. (2022). Long-range correlation analysis of high frequency prefrontal electroencephalogram oscillations for dynamic emotion recognition. *Biomed. Signal Process Control,* 72: 209.103291.

Gehrt, T.B., Macoveanu, J., Bailey, C.J., Fisher, P.M., Pallesen, K.J. and Frostholm, L. (2022). Resting-state connectivity and neural response to emotional images in patients with severe health anxiety: An MRI study. *Journal of Affective Disorders.*

Gentili, P.L. and Micheau, J. (2020). Light and chemical oscillations: review and perspectives. *Journal of Photochemistry and Photobiology C-photochemistry Reviews,* 43: 100321.

Ghandehari, O.O., Gallant, N.L., Hadjistavropoulos, T., Williams, J. and Clark, D.A. (2020). The relationship between the pain experience and emotion regulation in older adults. *Pain Medicine,* (12): 3366–3376.

Ghent, A. 2011. The happiness effect. *Bull World Health Organ.,* 89(4): 246–247.

Goldberg, S.B., Tucker, R.P., Greene, P.A., Davidson, R.J., Wampold, B.E. and Kearney, D.J. (2018). Mindfulness-based interventions for psychiatric disorders: A systematic review and meta-analysis. *Clin. Psychol. Rev.* (Feb.), 59: 52–60.

Goldenberg, A., Weisz, E., Sweeny, T.D., Cikara, M. and Gross, J.J. (2021). The crowd-emotion-amplification effect. *Psychological Science,* 32(3): 437–450.

González-Roldán, A.M., Bustan, S., Kamping, S., Flor, H. and Anton, F. (2023). Pain and related suffering reduce attention toward others. *Pain Practice: The Official Journal of World Institute of Pain,* 23(8): 873–885.

Grieder, M., Wang, D., Dierks, T., Wahlund, L.O. and Jann, K. (2018). Default mode network complexity and cognitive decline in mild AD's disease. *Front in Neurosci.,* 12: 770.

Grigg, G., Nowack, J., Bicudo, J., Bal, N.C., Woodward, H.N. and Seymour, R.S. (2021). Whole-body endothermy: ancient, homologous and widespread among the ancestors of mammals, birds and crocodylians. *Biological Reviews,* 97.

He, X. (2022). Relationship between self-esteem, interpersonal trust, and social anxiety of college students. *Occup. Ther. Int.,* 2022: 8088754.

Heelan, P.A. (1970). Complementarity, context dependence, and quantum logic. *Found. Phys.* 1: 95–110.

Herzog, M.H., Drissi-Daoudi, L. and Doerig, A. (2020). All in good time: long-lasting postdictive effects reveal discrete perception. *Trends in Cognitive Sciences,* 24: 826–837.

Hesp, C., Smith, R., Parr, T., Allen, M., Friston, K. and Ramstead, M.J. (2021). Deeply felt affect: the emergence of valence in deep active inference. *Neural Computation,* 33: 1–49.

Hooker, S.A., Masters, K.S. and Park, C. (2018). A meaningful life is a healthy life: a conceptual model linking meaning and meaning salience to health. *Review of General Psychology,* 22: 11–24.

Hopp, F.R., Amir, O., Fisher, J.T., Grafton, S.T., Sinnott-Armstrong, W. and Weber, R. (2023). Moral foundations elicit shared and dissociable cortical activation modulated by political ideology. *Nat. Hum. Behav.*

Hosseini Houripasand, M., Sabaghypour, S., Farkhondeh Tale Navi, F. and Nazari, M.A. (2023). Time distortions induced by high-arousing emotional compared to low-arousing neutral faces: an event-related potential study. *Psychological Research*, 1–12.

Hua, J.P., Trull, T., Merrill, A., Myers, O.T., Straub, K.T. and Kerns, J.G. (2020). Daily-life negative affect in emotional distress disorders associated with altered frontoinsular emotion regulation activation and cortical gyrification. *Cognitive Therapy and Research*, 1–18.

Huang, Y. (2019). Greater brain activity during the resting state and the control of activation during the performance of tasks. Sci. Rep., 9: 5027.

Huang, Y., Wu, R., Wu, J. (2020b). Psychological resilience, self-acceptance, perceived social support, and their associations with mental health of incarcerated offenders in China. *Asian J. Psychiatry*, 52: 102166.

Hummer, R.A. and Hernandez, E.M. (2013).The effect of educational attainment on adult mortality in the united states. *Popul. Bull.* (June) 68(1): 1–16.

Iacono, D., Markesbery, W.R., Gross, M., Pletnikova, O., Rudow, G. Zandi, P. and Troncoso, J.C. (2009). The nun study: clinically silent ad, neuronal hypertrophy, and linguistic skills in early life. *Neurology*, 73(9): 665–673.

Im, K., Lee, J., Lyttelton, O.C., Kim, S.H., Evans, A.C. and Kim, S. (2008). Brain size and cortical structure in the adult human brain. *Cerebral Cortex*, 18(9): 2181–2191.

Inagaki, T.K., Hazlett, L.I. and Andreescu, C. (2019). Naltrexone alters responses to social and physical warmth: Implications for social bonding. *Social Cognitive and Affective Neuroscience*, 14: 471 –479.

Inzlicht, M., Shenhav, A., Olivola, C.Y. (2018). The effort paradox: effort is both costly and valued. *Trends Cogn. Sci.*, 22(4): 337–349.

Ironside, M., Kumar, P., Kang, M.S. and Pizzagalli, D.A. (2018). Brain mechanisms mediating effects of stress on reward sensitivity. *Current Opinion in Behavioral Sciences*, 22: 106–113.

Izaki, T. and Ogawa, K. (2023). Dispersing attentional resources reduces negative emotions. *NeuroReport*.

Jaaro-Peled, H., Ayhan, Y., Pletnikov, M.V. and Sawa A. (2010). Review of pathological hallmarks of schizophrenia: comparison of genetic models with patients and nongenetic models. *Schizophr. Bull.* (Mar.), 36(2): 301–313.

Jamet, E., Gonthier, C., Cojean, S., Colliot, T. and Erhel, S. (2020). Does multitasking in the classroom affect learning outcomes? a naturalistic study. *Comput. Hum. Behav., 106*: 106264.

Jeffery, K.J. and Rovelli, C. (2020). Transitions in brain evolution: space, time and entropy. *Trends in Neurosciences*, 43: 467–474.

Johnson, S. (1999). *Who Moved My Cheese?* Vermilion.

Jung, Y.H., Shin, N.Y., Jang, J.H., Lee, W.J., Lee, D. and Choi, Y., (2019). Relationships among stress, emotional intelligence, cognitive intelligence, and cytokines. *Medicine (Baltimore),* (May), 98(18): e15345.

Justman, S. (2020). From blocked flows to suppressed emotions: the life of a trope. *Medical Humanities*, 48(1).

Kaiser, R.H., Whitfield, S., Gabrieli, D.G., Goer, F., Beltzer, M. and Minkel, J.D. (2016). Dynamic resting-state functional connectivity in major depression. *Neuropsychopharmacology*, 41(7): 1822–1830.

Kao, F-C., Wang, SR. and Chang, Yj. (2015). Brainwaves analysis of positive and negative emotions. *ISAA*, 12: 1263–1266.

Kataoka, N., Shima, Y., Nakajima, K. and Nakamura, K. (2020). A central master driver of psychosocial stress responses in the rat. *Science*, 367: 1105–1112.

Kearney, M.S. and Levine, P.B. (2014). Income inequality, social mobility, and the decision to drop out of high school. *Brookings Papers on Economic Activity*, 2016, 333–396.

Khrennikov, A. (2020). Quantum-like modeling: cognition, decision-making, and rationality. *Mind Soc.,* 19: 307–310.

Khrennikov, A. (2020). Social laser model for the bandwagon effect: generation of coherent information waves. *Entropy,* 22 (5): 559.

Khrennikov, A. (2023). Coherent decision making stimulated within the social laser: open quantum systems framework. *Phil. Trans. R. Soc. A.,* 3812022029420220294.

Kirk, J.M., Doty, P. and De Wit, H. (1998). Effects of expectancies on subjective responses to oral delta9-tetrahydrocannabinol. *Pharmacol. Biochem. Behav.,* 59: 287–293.

Kleiner, T.-M. (2018). Public opinion polarisation and protest behaviour. *European Journal of Political Research,* 57: 941–962.

Klimek, P., Poledna, S. and Thurner, S. (2019). Quantifying economic resilience from input–output susceptibility to improve predictions of economic growth and recovery. *Nat. Commun.,* 10: 1677.

Knapen, T. (2021). Topographic connectivity reveals task-dependent retinotopic processing throughout the human brain. *PNAS,* 118(2): e2017032118.

Knight, E.L., Sarkar, A., Prasad, S. and Mehta, P.H. (2019). Beyond the challenge hypothesis: The emergence of the dual-hormone hypothesis and recommendations for future research. *Hormones and Behavior,* 123.

Koban, L., Gianaros, P.J., Kober, H. and Wager, T.D. (2021). The self in context: brain systems linking mental and physical health. *Nat. Rev. Neurosci.,* 22: 309–322.

Kolvoort, I.R., Wainio-Theberge, S., Wolff, A. and Northoff, G. (2020). Temporal integration as "common currency" of brain and self-scale-free activity in resting-state EEG correlates with temporal delay effects on self-relatedness. *Human Brain Mapping,* 41(15): 4355–4374.

Kong, F., Gong, X., Sajjad, S., Yang, K. and Zhao, J. (2019). How is emotional intelligence linked to life satisfaction? the mediating role of social support, positive affect and negative affect. *Journal of Happiness Studies,* 20: 2733–2745.

Kong, Y. (2022). Are emotions contagious? a conceptual review of studies in language education. *Front Psychol.,* 13:1048105.

Kool, W. and Botvinick, M. (2018). Mental labour. *Nat. Hum. Behav.,* 2: 899–908.

Koonin, E.V. (2016). The meaning of biological information. *Philosophical Transactions. Series A, Mathematical, Physical, and Engineering Sciences,* 374(2063): 20150065.

Kostic, M. (2014). The elusive nature of entropy and its physical meaning. *Entropy,* 16: 953–967.

Krems, J.A., Hahnel-Peeters, R.K., Merrie, L.A., Williams, K.E. and Sznycer, D. (2023). Sometimes we want vicious friends: People have nuanced preferences for how they want their friends to behave toward them versus others. *Evolution and Human Behavior,* 44(2): 88–98.

Krause, M.S. (1972). An analysis of Festinger's Cognitive dissonance theory. Philosophy of Science, 39: 32–50.

Kropotkin, P. (1902). *Mutual Aid: A Factor of Evolution. Extending Horizons.* Columbia University.

Kuno, Y. (1956). *Human Perspiration.* Springfield, IL: Charles C Thomas. https://journals.sagepub.com/doi/full/10.1177/1094428116681073#

Levine, E., Bitterly, T.B., Cohen, T.R. and Schweitzer, M.E. (2018). Who is trustworthy? Predicting trustworthy intentions and behavior. *Journal of Personality and Social Psychology,* 115(3): 468–494.

Li, X., Zhu, Z., Zhao, W., Sun, Y., Wen, D. and Xie, Y. (2018). Decreased resting-state brain signal complexity in patients with mild cognitive impairment and AD's disease: A multiscale entropy analysis. *Biomed. Opt. Express,* 9(4): 1916–1929.

Libet, B. (1985). Unconscious cerebral initiative and the role of conscious will in voluntary action. *The Behavioral and Brain Sciences*, 8: 529–566.

Lin, L., Chang, D., Song, D., Li, Y. and Wang, Z. (2022). Lower resting brain entropy is associated with stronger task activation and deactivation. *NeuroImage*, 249.

Lucia, U. and Grisolia, G. (2020). Time and clocks: a thermodynamic approach. *Results in Physics*, 16: 102977.

Lupien, S.J., Maheu, F., Tu, M., Fiocco, A. and Schramek, T.E. (2007). The effects of stress and stress hormones on human cognition: Implications for the field of brain and cognition. *Brain Cogn.*, 65(3): 209–37.

MacCann, C., Jiang, Y., Brown, L.E., Double, K.S., Bucich, M. and Minbashian, A. (2019). Emotional intelligence predicts academic performance: a meta-analysis. *Psychological Bulletin*.

Makarevskaya, Y.E. (2018). Specificity of the structure of the intellect of socially successful people. *Psychology and Education – ICPE*. 10.15405/epsbs.2018.11.02.42.

Makharia, A., Nagarajan, A.S. and Mishra, A. (2016). Effect of environmental factors on intelligence quotient of children. *Industrial Psychiatry Journal*, 25:189–194.

Mani, A., Mullainathan, S., Shafir, E. and Zhao, J. (2013). Poverty impedes cognitive function. *Science*, 341: 976–980.

Manohar, S.G., Muhammed, K., Fallon, S.J. and Husain, M. (2018). Motivation dynamically increases noise resistance by internal feedback during movement. *Neuropsychologia*, 1–11.

Martyushev, L.M. (2013). Entropy and entropy production: old misconceptions and new breakthroughs. *Entropy*, 15: 1152–1170.

Mattarozzi, K., Bagnis, A. and Kłosowska, J. (2023). No(cebo) vax: COVID-19 vaccine beliefs are important determinants of both occurrence and perceived severity of common vaccines' adverse effects. *Psychological Science, 34*: 603–615.

Meeusen, R., Van Cutsem, J. and Roelands, B. (2020). Endurance exercise-induced and mental fatigue and the brain. *Experimental Psychology*, 106(12).

Mehta, P.H. and Josephs, R.A. (2010). Testosterone and cortisol jointly regulate dominance: Evidence for a dual-hormone hypothesis. *Hormones and Behavior*, 58: 898–906.

Meijers, J., Harte, J. M., Meynen, G., Cuijpers, P. and Scherder, E.J. (2018). Reduced self-control after three months of imprisonment: a pilot study. *Front. Psychol.*, 9: 69.

Miller, A. (1949). *Death of a Salesman*. Viking.

Moffitt, T.E., Arseneault, L., Belsky, D. (2011). A gradient of childhood self-control predicts health, wealth, and public safety. *PNAS*, 108(7): 2693–8.

Mullen, E. and Skitka, L.J. (2006). Exploring the psychological underpinnings of the moral mandate effect: Motivated reasoning, group differentiation, or anger? *Journal of Personality and Social Psychology*, 90(4): 629–643.

Murav'eva, E.D. and Shelepin, Yu. (2022). Fractal structure of brain electrical activity of patients with mental disorders. *Frontiers in Physiology*, 13.

Nazmi Sofian Suhaimi, James Mountstephens, Jason, Teo. (2020). EEG-Based emotion recognition: a state-of-the-art review of current trends and opportunities. *Computational Intelligence and Neuroscience*, vol. Article ID 8875426, p. 19.

Neal, D.T., Wood, W., Wu, M. and Kurlander, D. (2011). The pull of the past: when do habits persist despite conflict with motives? *Pers. Soc. Psychol. Bull.*, 37(11):1428–1437.

Nicolas, M., Martinent, G., Millet, G., Bagneux, V. and Gaudino, M. (2019). Time courses of emotions experienced after a mountain ultra-marathon: Does emotional intelligence matter? *Journal of Sports Sciences*, 37: 1831–1839.

Northoff, G., Wainio-Theberge, S. and Evers, K. (2019). Is temporo-spatial dynamics the "common currency" of brain and mind? in quest of "spatiotemporal neuroscience". *Physics of Life Reviews*.

Nuno-Perez, A., Trusel, M. and Mameli, M. (2021). Stress undermines reward-guided cognitive performance through synaptic depression in the lateral habenula. *Neuron*, 109(6): 947–956.

Ohanian, H.C. (1986).   HYPERLINK. "https://physics.mcmaster.ca/phys3mm3/notes/whatisspin.pdf" "What is spin?" *American Journal of Physics*, 54(6): 500–505.

O'Neill, J. and Schoth, A. (2022). The mental maxwell relations: a thermodynamic allegory for higher brain functions. *Frontiers in Neuroscience*, 16: 827888.

Padamsey, Z., Katsanevaki, D., Dupuy, N. and Rochefort, N.L. (2022). Neocortex saves energy by reducing coding precision during food scarcity. *Neuron*, 110: 280–296. e10.

Pang, J.C., Aquino, K.M., Oldehinkel, M., Robinson, P.A., Fulcher, B.D., Breakspear, M. and Fornito, A. (2023). Geometric constraints on human brain function. *Nature*.

Peng, L. and Xie, T. (2016). Making similarity versus difference comparison affects perceptions after bicultural exposure and consumer reactions to culturally mixed products. *Journal of Cross-Cultural Psychology*, 47(10): 1380–1394.

Pepperell, R. (2018) Consciousness as a physical process caused by the organization of energy in the brain. *Frontiers Psychol.*, 9.

Perl, O., Shuster, A., Heflin, M., Na, S., Kidwai, A., Booker, N. et al. (2023). A thalamic circuit represents dose-like responses induced by nicotine-related beliefs in human smokers. bioRxiv.

Pinto, Y., Neville, D.A., Otten, M., Corballis, P.M., Lamme, V.A. and de Haan, E.H. (2017). Split brain: Divided perception but undivided consciousness. *Brain*, 140(5): 1231–1237.

Piper, Watty, 1930. The Little Engine That Could. *Grosset and Dunlap.*

Plato, 375 BC. Republic, Les Prairies Numeriques.

Plomin, R. and von Stumm, S. (2018). The new genetics of intelligence. *Nat. Rev. Genet.* (Mar.), 19(3): 148–159.

Pothos, E.M. and Busemeyer, J.R. (2022). Quantum Cognition. *Annual Review of Psychology*, 73: 749–778.

Preston, J.L., Coleman, T.J. and Shin, F. (2023). Spirituality of science: implications for meaning, well-being, and learning. *Personality and Social Psychology Bulletin*, 0(0).

Protzko, J. (2016).   Does the raising IQ-raising g distinction explain the fadeout effect? *Intelligence*, 56: 65–71.

Raison, C.L., Hale, M.W., Williams, L.E., Wager, T.D. and Lowry, C.A. (2015). Somatic influences on subjective well-being and affective disorders: The convergence of thermosensory and central serotonergic systems. *Front. Psychol.*, 5,1589.

Remmers, C. and Zander, T. (2018). Why you don't see the forest for the trees when you are anxious: anxiety impairs intuitive decision making. *Clin. Psychol. Sci.*, 6: 48–62.

Rigney, D. (2010). *"Matthew Effects in the Economy." The Matthew Effect: How Advantage Begets Further Advantage*. Columbia University Press, pp. 35–52.

Roberts, B.W. (2019). Time Reversal. URL: http://philsci-archive.pitt.edu/id/eprint/15033.

Roemer, E. and Borkovec, T.D. (1994). Effects of suppressing thoughts about emotional material. *Journal of Abnormal Psychology*, 103: 467–474.

Rossi, R. (2021). The role of introspective evaluation of intentions to act in a quantum-like cognitive model. *Physics Letters A*, 390: 127111.

Rothenhoefer, K.M., Hong, T., Alikaya, A. and Stauffer, W.R. (2021). Rare rewards amplify dopamine responses. *Nat. Neurosci.*, 24: 465–469.

Rousson, A.N., Fleming, C.B. and Herrenkohl, T.I. (2020).   Childhood maltreatment and later stressful life events as predictors of depression: A test of the stress sensitization hypothesis. *Psychol. Violence* (Sept.),;10(5): 493–500.

Ruan, Y., Reis, H., Zareba, W. and Lane, R. (2020). Does suppressing negative emotion impair subsequent emotions? Two experience sampling studies. *Motivation and Emotion*, 44: 427–435.

Rubinstein, J.S., Meyer D.E. and Evans J.E. (2001). Executive control of cognitive processes in task switching. *J. Exp. Psychol. Hum. Percept. Perform.*, 27(4): 763–797.

Rudd, M., Vohs, K.D. and Aaker, J.L. (2012). Awe expands people's perception of time, alters decision making, and enhances well-being. *Psychological Science*, 23:1130–1136.

Ryan, R.M., and Deci, E.L. (2017). *Self-Determination Theory: Basic Psychological Needs in Motivation Development and Wellness.* New York, NY: Guilford Press.

Saaty, T.L. and Vargas, L.G. (2017). Origin of neural firing and synthesis in making comparisons. *European Journal of Pure and Applied Mathematics*, 10(4),: 602–613.

Sakai, J.A. (2020). Core concept: how synaptic pruning shapes neural wiring during development and, possibly, in disease. *Proceedings of the National Academy of Sciences*, 117: 16096–16099.

Saunders, B., Lin, H., Milyavskaya, M. and Inzlicht, M. (2017). The emotive nature of conflict monitoring in the medial prefrontal cortex. *International Journal of Psychophysiology*, 119: 31–40.

Savage, L.J. (1954). *The Foundations of Statistics.* John Wiley and Sons.

Scheidel, W. (2017). *The Great Leveler, Violence and the History of Inequality from the Stone Age to the Twenty-first Century.* Pp. 192–198. Princeton, Oxford.

Schoeller, F., Perlovsky, L. and Arseniev, D. (2018). Physics of mind: experimental confirmations of theoretical predictions. *Phys. Life Rev.* (Aug.), 25: 45–68.

Schubert, T., Eloo, R., Scharfen, J. and Morina, N. (2019). How imagining personal future scenarios influences affect: systematic review and meta-analysis. *Clinical Psychology Review*, 75: 101811.

Seif, A., Hafezi, M. and Jarzynski, C. (2021). Machine learning the thermodynamic arrow of 13 time. *Nature Physics*, 17(1): 105–113.

Selesnick, S.A. and Piccinini, G. (2018). Quantum-like behavior without quantum physics II. a quantum-like model of neural network dynamics. *Journal of Biological Physics*, 44: 501–538.

Shannon, C.E. (1948). A mathematical theory of communication. *Bell. Syst. Tech. J.*, 27: 379–423.

Shi, L., Beaty, R.E., Chen, Q., Sun, J., Wei, D., Yang, W. and Qiu, J. (2019). Brain entropy is associated with divergent thinking. *Cerebral Cortex* (Mar.), 30(2): 708–717.

Shimazaki, H. (2020). The principles of adaptation in organisms and machines II: thermodynamics of the bayesian brain. arXiv: *Neurons and Cognition*.

Sizemore, A.E., Giusti, C., Kahn, A., Vettel, J.M., Betzel, R.F., and Bassett, D.S. (2018). Cliques and cavities in the human connectome. *J. Comput. Neurosci.*, 44: 115.

Smallwood, J., Bernhardt, B.C., Leech, R., Bzdok, D., Jefferies, E. and Margulies, D.S. (2021). The default mode network in cognition: A topographical perspective. *Nat. Rev. Neurosci.*, 22: 503–513.

Smitha, K.A., Akhil Raja, K., Arun, K.M., Rajesh, P., Thomas, B., Kapilamoorthy, T. and Kesavadas, C. (2017). Resting state fMRI: A review on methods in resting state connectivity analysis and resting state networks. *Neuroradiol. J.*, 4: 305–317.

Sridhar, V.H., Li, L., Gorbonos, D., Schell, B.R., Sorochkin, T., Gov, N.S. and Couzin, I.D. (2021). The geometry of decision-making in individuals and collectives. *Proceedings of the National Academy of Sciences of the United States of America*, 118.

Starkweather, C.K., Gershman, S.J. and Uchida, N. (2018). The Medial Prefrontal Cortex shapes dopamine reward prediction errors under State Uncertainty. *Neuron*, 98(3): 616–629.e6.

Stewart, A.J. and Plotkin, J.B. (2013). From extortion to generosity, evolution in the iterated prisoner's dilemma. *Proceedings of the National Academy of Sciences*, 110: 15348–15353.

Stringer, C., Pachitariu, M., Steinmetz, N., Carandini, M. and Harris, K.D. (2019). High-dimensional geometry of population responses in visual cortex. *Nature,* 571: 361–365.

Surov, I.A. (2022). Natural code of subjective experience. *Biosemiotics,* 15: 109–139.

Tajadura-Jiménez, A., Banakou, D., Bianchi-Berthouze, N. and Slater, M. (2017). Embodiment in a child-like talking virtual body influences object size perception, self-identification, and subsequent real speaking. *Sci. Rep.* 7: 9637.

Tambini, A. and Davachi, L. (2019). Awake reactivation of prior experiences consolidates memories and biases cognition. *Trends in Cognitive Sciences,* 23: 876–890.

Tesser, A. (1988). Toward a self-evaluation maintenance model of social behavior. *Advances in Experimental Social Psychology,* 21: 181–227.

Tkadlec, J., Pavlogiannis, A., Chatterjee, K. and Nowak, M.A. (2020). Limits on amplifiers of natural selection under death-birth updating. *PLOS Computational Biology,* 16(1): e1007494.

Tong, F., Meng, M. and Blake, R. (2006). Neural bases of binocular rivalry. *Trends Cogn. Sci.,* 10: 502–511.

Tooley, K.L. (2020). Effects of the human gut microbiota on cognitive performance, brain structure and function: a narrative review. *Nutrients (Sept.),* 12(10): 3009.

Tozzi, A., Peters, J.F., Fingelkurts, A.A., Fingelkurts, A.A. and Marijuán, P.C. (2017). Topodynamics of metastable brains. *Phy. Life Rev.* (July), 21: 1–20.

Treadway, M.T., Woodward, N.D., Li, R., Ansari, M. and Baldwin, R.M. (2012). Dopaminergic mechanisms of individual differences in human effort-based decision-making. *J Neurosci.,* 32(18): 6170–6176.

Trevisiol, A. (2017). Monitoring ATP dynamics in electrically active white matter tracts. eLife.

Trezza, V., Baarendse, P.J. and Vanderschuren, LJ. (2010). The pleasures of play: pharmacological insights into social reward mechanisms. *Trends Pharmacol. Sci.* (Oct.), 31(10): 463–469.

Tring, E., Dipoppa, M. and Ringach, D.L. (2023). A power law of cortical adaptation. bioRxiv.

Tsao, A., Sugar, J., Lu, L., Wang, C., Knierim, J.J., Moser, M. and Moser, E.I. (2018). Integrating time from experience in the lateral entorhinal cortex. *Nature,* 561(7721): 57–62.

Türker, B., Musat, E.M., Chabani, E. Fonteix-Galet, A., Maranci, J.B. and Wattiez, N. (2023). Behavioral and brain responses to verbal stimuli reveal transient periods of cognitive integration of the external world during sleep. *Nat Neurosci.,* 26: 1981–1993.

Tversky A. and Shafir, E. (1992). The disjunction effect in choice under uncertainty. *Psychol. Sci.,* 3: 305–309.

Tversky, A. and Griffin, D. (1991). Endowment and judgment in contrast of well-being. *In:* Strack, F., Argyle, M. and Schwarz, N. (Eds.), *Subjective Well-Being: An Interdisciplinary Perspective.* Oxford UK: Pergamon.

Uhrig, M., Trautmann, N., Baumgärtner, U., Treede, R., Henrich, F., Hiller, W. et al. (2016). Emotion elicitation: a comparison of pictures and films. *Frontiers in Psychology,* 7, Article 180.

Umbach, R., Raine, A. and Leonard, N.R. (2018). Cognitive decline as a result of incarceration and the effects of a CBT/MT intervention: a cluster-randomized controlled trial. *Criminal Justice and Behavior,* 45: 31–55.

Varley, T.F., Denny, V., Sporns, O. and Patania, A. (2021). Topological analysis of differential effects of ketamine and propofol anaesthesia on brain dynamics. *Royal Society Open Science,* 8(6): 201971.

Velasco, G.G., Fernández, T., Silva-Pereyra, J. and Alcántara, V.R. (2019). Higher cognitive reserve is associated with better neural efficiency in the cognitive performance of young adults. An event-related potential study. bioRxiv.

Volkow, N.D., Wang, G., Ma, Y., Fowler, J.S., Zhu, W. and Maynard, L. (2003). Expectation enhances the regional brain metabolic and the reinforcing effects of stimulants in cocaine abusers. *J. Neurosci.* 23: 11461–11468.

Vries, M.F. (2017). Down the rabbit hole of shame. INSEAD Working Paper Series.

Wade-Bohleber, L.M., Boeker, H., Grimm, S., Gärtner, M., Ernst, J. and Recher, D. (2020). Depression is associated with hyperconnectivity of an introspective socio-affective network during the recall of formative relationship episodes. *J. Affect. Disord.*, 274: 522–534.

Wang, J. and Lapate, R.C. (2023). Emotional state dynamics impacts temporal memory. bioRxiv.

Wang, Z. (2020) Brain entropy mapping in healthy aging and AD's disease. *Front. Aging Neurosci.*, 12: 596122.

Wang, Z. (2021). The neurocognitive correlates of brain entropy estimated by resting state fMRI. *NeuroImage*, 232.

Wato, W., Kochiyama, T., Uono, S., Sawada, R. and Yoshikawa, S. (2020) Amygdala activity related to perceived social support. *Sci. Rep.*, 10: 2951.

Watts, M.E., Pocock, R. and Claudianos, C. (2018). Brain energy and oxygen metabolism: emerging role in normal function and disease. *Frontiers in Molecular Neuroscience*, 11.

Watts, T.W., Duncan, G.J. and Quan, H. (2018). Revisiting the marshmallow test: a conceptual replication investigating links between early delay of gratification and later outcomes. *Psychological Science*, 29(7).

Wendt, A. (2015). *Quantum Mind and Social Science. Unifying Physical and Social Science. Unifying Physical and Social Ontology.* Cambridge: Cambridge University Press.

Wenzlaff, R.M. and Wegner, D.M. (2000). Thought suppression. *Annual Review of Psychology*, 51: 59–91.

Wissner-Gross, A.D. and Freer, C.E. (2013). Causal entropic forces. *Physical Review Letters*, 110 (16): 168702.

Wolff, A., Giovanni, D.A., Gomez-Pilar, J., Nakao, T., Huang, Z., Longtin, A. and Northoff, G. (2019). The temporal signature of self: Temporal measures of resting-state EEG predict self-consciousness. *Human Brain Mapping*, 40: 789–803.

Xu, AJ. and Schwarz, N. (2018). How one thing leads to another: spillover effects of behavioral mindsets. *Current Directions in Psychological Science*, 27(1): 51–55.

Xu, S., Zhang, Z., Li, L., Zhou, Y., Lin, D. and Zhang, M. (2023). Functional connectivity profiles of the default mode and visual networks reflect temporal accumulative effects of sustained naturalistic emotional experience. *NeuroImage*.

Yang, J., Zhao, Q., Zhao, X., Wu, D., Li, M. and Zhang, W. (2020). User's attitude under the perspective of mental energy flow. *Mathematical Problems in Engineering*, 2020: 1–14.

Yang, S., Zhao, Z. and Cui, H. (2019). Temporal variability of cortical gyral-sulcal resting state functional activity correlates with fluid intelligence. *Frontiers in Neural Circuits*, 13: 36.

Yang, S., Zhao, Z., Cui, H., Zhang, T., Zhao, L. and He, Z. (2019). Temporal variability of cortical gyral-sulcal resting state functional activity correlates with fluid intelligence. Frontiers in Neural Circuits, 13.

Yuksel, C., Denis, D., Coleman, J., Oh, A., Cox, R. and Morgan, A. (2023). Emotional memories are enhanced when reactivated in slow-wave sleep but impaired in REM. *bioRxiv*.

Zanin, M., Güntekin, B., Aktürk, T., Hanoglu, L. and Papo, D. (2020). Time irreversibility of resting-state activity in the healthy brain and pathology. *Frontiers in Physiology*, 10.

Zappasodi, F., Marzetti, L., Olejarczyk, E., Tecchio, F. and Pizzella, V. (2015). Age-related changes in electroencephalographic signal complexity. *PLoS ONE*, 10.

Zeldetz, V., Natanel, D. and Boyko, M. (2018). A new method for inducing a depression-like behavior in rats. *J. Vis. Exp.*, 132: 57137.

Zhang, Q., Sun, S., Zheng, X. and Liu, W. (2019). The role of cynicism and personal traits in the organizational political climate and sustainable creativity. *Sustainability*, 11: 257.

Zhou, W. and Chen, D. (2009). Binaral rivalry between the nostrils and in the cortex. *Current Biology*, 19: 1561–1565.

Zmigrod, L., Zmigrod, S., Rentfrow, P.J. and Robbins, T. (2019). The psychological roots of intellectual humility: The role of intelligence and cognitive flexibility. *Personality and Individual Differences*, 141: 200–208.

Index

For Product Safety Concerns and Information please contact our EU
representative GPSR@taylorandfrancis.com
Taylor & Francis Verlag GmbH, Kaufingerstraße 24, 80331 München, Germany